Environmental Modelling with GIS and Remote Sensing

Edited by Andrew Skidmore

London and New York

First published 2002
by Taylor & Francis
11 New Fetter Lane, London EC4P 4EE

Simultaneously published in the USA and Canada
by Taylor & Francis Inc,
29 West 35th Street, New York, NY 10001

Taylor & Francis is an imprint of the Taylor & Francis Group

Printed and bound in Great Britain by
Biddles Ltd, Guildford & King's Lynn

British Library Cataloguing in Publication Data
A catalogue record for this book is available from the British Library

Library of Congress Cataloging in Publication Data
A catalogue record has been requested

ISBN 0–415–24170–7

Contents

Preface

This book is the summary of lectures presented at a short course entitled "Environmental Modelling and GIS" at the International Institute for Aerospace Survey (ITC), The Netherlands. Previous books on environmental modelling and GIS are detailed in Chapter 1. This book aims to bring the literature up to date, as well as provide new perspectives on developments in environmental modelling from a GIS viewpoint.

Environmental modelling remains a daunting task – decision makers, politicians and the general public demand faster and more detailed analyses of environmental problems and processes, and clamour for scientists to provide solutions to these problems. For GIS users and modellers, the problems are multi-faceted, ranging from access to data, data quality, developing and applying models, as well as institutional and staffing issues. These topics are covered within the book. But the main emphasis of the book is on environmental models; a good overview of currently available data, models and approaches is provided.

There is always difficulty in developing a coherent book from submitted chapters. We have tried to ensure coherence through authors refereeing each other's chapters, by cross-references, by indexing, and finally by editorial input. Ultimately, it was not possible to rewrite every chapter into a similar style – it would destroy the unique contribution of authors of each chapter. And the editor would overstep the bounds of editorship and drift into authorship.

It is assumed that the reader has basic knowledge about GIS and remote sensing, though most chapters are accessible to beginners. An introductory text for GIS is Burrough and McDonnell (1993) and for remote sensing Avery and Berlin (1992).

The editor and authors would like to acknowledge the assistance of the following:

Daniela Semeraro who helped organize the short course, and provided secretarial services during the production of the book. **Gulsaran Inan** who assisted in completing the book.

The **participants** of the course who gave feedback and comments.

ITC management (Professor Karl Harmsen and Professor Martien Molenaar) for facilitating the short course and providing staff time to produce the book.

The assistance of **ITC staff** in running the short course.

The **host organizations** (see affiliations) of the authors who provided staff with the time to write the chapters.

The **Taylor and Francis staff** (especially Tony Moore and Sarah Kramer) who supported the editor and authors during the production of the book.

REFERENCES CITED

Avery, T. E. and Berlin G. L. (1992). *Fundamentals of Remote Sensing and Airphoto Interpretation*. New York, Macmillan Publishing Company.
Burrough, P. A. and McDonnell, R.A. (1993). *Principles of Geographical Information Systems*. Oxford, Clarendon Press.

Andrew K. Skidmore
ITC, Enschede, The Netherlands
November 2001

List of Figures

List of Tables and Boxes

xviii *Tables and Boxes*

1

Introduction

Andrew K. Skidmore

1.1 THE CHALLENGE

The environment is key to sustaining human economic activity and well-being, for without a healthy environment, human quality of life is reduced. Most people would agree that there are also many reasons to protect the environment for its own inherent worth, and especially to leave a legacy of fully functioning natural resources. Sustainable land management refers to the activities of humans and implies that activity will continue in perpetuity. It is a term that attempts to balance the often conflicting ideals of economic growth and maintaining environmental quality and viability.

There are three interacting components required for successful natural resource and environmental management, namely policy, participation and information (Figure 1.1). These factors are especially critical in less developed countries, where infrastructure is often rudimentary. The balance between these three components, and their influence on management, will depend on the management problem, as well as the infrastructure and the social, economic and cultural traditions of the country.

Sustainable land use and development is based on two critical factors. Firstly, national, regional and local policy and leadership, which may be asserted through diverse mechanisms including legislation, policy documents, imposing sanctions, introducing incentives (reduced tax, subsidies, etc.), motivation to contribute to development and so on. Policy tools are necessary to encourage farmers and other natural resource managers to make good use of natural resources, and organize management in a sustainable manner. Policy may also be used to define protection areas. Secondly, sustainable land use requires the participation by, and benefits to, local people (farmers, managers, land owners, stakeholders). If the local people benefit directly (through an improved standard of living, better environment, gender equality, etc.) then they will contribute positively to the policy settings. In addition, an active non-governmental organization network is often effective in maintaining accountability.

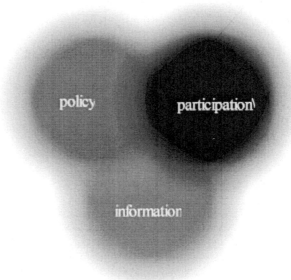

Figure 1.1: Ingredients necessary for a successful geographic information system (GIS) for environmental management project – policy, participation and information.

These elements require the provision of timely, accurate and detailed information on land resources as well as changes in the land resource. This spatial information is provided, and applied through, a geographic information system (GIS) and remote sensing. Better spatial information and maps leads to improved planning and decision making at all levels and scales, and hopefully generates harmony between production and conservation across a landscape.

This book is focused on information, and specifically on how spatial information may be used for environmental modelling and management. There are also examples and discussion in the following chapters of how the geographical information is used for policy, as well as for participation and planning. Thus, the book emphasizes environmental models, and specifically how these models have been extrapolated in space and time in order to develop policy and ultimately assist managers of the environment.

1.2   MOTIVATION TO WRITE THIS BOOK

The genesis for this book was a one week commercial short course developed at ITC (International Institute for Aerospace Survey and Earth Sciences) for participants from around the world. It was an opportunity for a diverse group of lecturers to present the current state of knowledge in their field of expertise, and to share and comment on developments in related fields. The participants and lecturers came from all regions of the World (Europe, Africa, Asia, the Pacific, and the Americas) (Table 1.1). The participants could be broadly categorized as mid-career professionals, some occupying senior management positions in both private and government organizations. All were concerned with the management and implementation of GIS and remote sensing, and specifically how environmental models can be put to work for environmental management. Their excellent feedback at the end of the course allowed us to write a short chapter on perceived trends and bottlenecks in the field of GIS and environmental modelling.

Table 1.1: Participants and lecturers in the short course.

Region	Number of participants and lecturers
Africa	10
Asia	6
Americas	4
Europe	14
Oceania	3

The other motivation for writing this edited volume is that there are few books that bring together case studies, theory and applications of GIS and environmental modelling in one volume. Even though we attempt to present the state of the art, GIS and remote sensing is continuing to develop quickly, so some elements of the book may rapidly become dated.

This book is aimed at GIS and remote sensing professionals. We anticipate it will also be of interest to university students, either as the basis of short courses, or as a semester course on the environmental application of GIS or remote sensing. The book complements texts on the introduction to GIS (e.g. Burrough 1986; Burrough and McDonnell 1993), management of GIS (e.g. Aronoff 1989; Huxhold and Levinsohn 1995), the application of GIS (e.g. Heit and Shortreid, 1991), advanced data structures (e.g. Laurini and Thompson 1992; Van Oosterom 1993) and an earlier book on environmental modelling with GIS (Goodchild *et al.* 1993).

1.3 WHAT IS ENVIRONMENTAL MODELLING AND HOW CAN GIS AND REMOTE SENSING HELP IN ENVIRONMENTAL MODELLING?

A model is an abstraction or simplification of reality (Odum 1975; Jeffers 1978; Duerr *et al.* 1979). When models are applied to the environment, it is anticipated that insights about the physical, biological or socio-economic system may be derived. Models may also allow prediction and simulation of future conditions, both in space and in time. The reason to build models is to understand, and ultimately manage, a sustainable system.

As introduced in section 1.1, sustainable development has been defined in many ways; in fact there are 67 different definitions listed in the 'Introduction to natural resource management' module taught at ITC (McCall, per comm.). Interestingly, none mention GIS and remote sensing as being tools necessary for sustainable development. Sustainable development is a term that attempts to balance the often conflicting ideals of economic growth while maintaining environmental quality and viability. As such, sustainability implies maintaining components of the natural environment over time (such as biological diversity, water quality, preventing soil degradation), while simultaneously maintaining (or improving) human welfare (e.g. provision of food, housing, sanitation, etc.).

In any definition of sustainability, a key element is change; for example, Fresco and Kroonenberg (1992) define sustainability as the "...dynamic equilibrium between input and output". In other words, they emphasize that dynamic equilibrium implies change and that in order for a land system to be sustainable, its potential for production should not decrease (in other words the definition allows for reversible damage). This type of definition is most applicable when considering agricultural production systems, but may also be generalized to the management of natural areas. A broader definition of sustainability includes the persistence of all components of the biosphere, even those with no apparent benefit to society, and relates particularly towards maintaining natural ecosystems. Other definitions emphasize increasing the welfare of people (specifically the poor at the 'grassroots' level) while minimizing environmental damage (Barbier 1987), which has a socio-economic bias. The necessary transition to renewable resources, is emphasised by Goodland and Ledec (1987) who state that renewable resources should be used in a way which does not degrade them, and that non-renewable resources should be used so that they allow an orderly societal transition to renewable energy sources.

That changes continually occur at many spatial (e.g. global, regional, local) and temporal (e.g. ice ages, deforestation, fire) scales is obvious to any observer. For example, change may occur in the species occupying a site, amount of nitrate in ground water, or crop yield from a field. Change may also occur to human welfare indices such as health or education. To assess whether such changes are sustainable is a non-trivial problem. Possibly the advantage of the debate about sustainable development is that the long-term capacity of the earth to maintain human life through a healthy and properly functioning global ecosystem is now a normal political goal.

Many alternative definitions of GIS have been suggested, but a simple definition is that a GIS is a computer-based system for the capture, storage, retrieval, analysis and display of spatial data. GIS are differentiated from other spatially related systems by their analytical capacity, thus making it possible to perform modelling operations on the spatial data (see Box 1.1). The spatial data in GIS databases are predominately generated from remote sensing through the direct import of images and classified images, but also through the generation of conventional maps (e.g. topographic maps) using photogrammetry. Thus remote sensing is an integral part of GIS, and GIS is impossible without remote sensing. Remote sensing data, such as satellite images and aerial photos allow us to map the variation in terrain properties, such as vegetation, water, and geology, both in space and time. Satellite images give a synoptic overview and provide very useful environmental information for a wide range of scales, from entire continents to details of a metre.

Box 1.1: Typical questions addressed by a GIS (after Burrough 1986 and Garner 1993).

Location {What is at ....?}
– What house number is at lot X?
– What is the crop type at point Y?

Condition {Where is it .....?}
– Where is all the land zoned for firework factories within 200 m of a suburb?
– Where is the forest within 100 km of a timber mill?

Distribution {What is the distribution/pattern .....?}
– What proportion of *Eucalyptus sieberi* trees occur on ridges?
– What is the average radiation level 5 – 10 km from Chernobyl?

Trend {What has changed ......?}
– What is the decline in wildebeest abundance in the last 20 years in Kenya?
– What is the increase in salt in the Murray River in the last 50 years?

Routing {Which is the best way ....?}
– Which is the shortest distance to a forest fire?
– Which is the fastest route from London to Colchester?

It is because of its analytical capacity that GIS is being increasingly used for decision making, planning and environmental management. GIS and remote sensing have been combined with environmental models for many applications, including for example, monitoring of deforestation, agro-ecological zonation, ozone layer depletion, food early warning systems, monitoring of large atmospheric-oceanic anomalies such as El Nino, climate and weather prediction, ocean mapping and monitoring, wetland degradation, vegetation mapping, soil mapping, natural disaster and hazard assessment and mapping, and land cover maps for input to global climate models. Though developments have been broadly based across many divergent disciplines, there is still much work required to develop GIS models suited to natural resource management, refine techniques, improve the accuracy of output, and demonstrate and implement work in operational systems.

In the mainstream GIS literature there has been growing attention to the need to consider models as an integral part of GIS, and to improve understanding and application of models. The problems inherent in applying GIS (for environmental management) are summarized in Table 1.2, as identified by key papers and books. An obvious trend in Table 1.2 is for technical problems (hardware, software, etc.) in the 1980s to be replaced by organizational issues in the 1990s. Data availability, quality of data, and appropriate models for specific applications appear to be the main issues presently, and are the focus in this book.

Table 1.2: Main problems in implementing GIS, as cited from 1986 to 2001.

Year	Author	Problems in implementing GIS
1986	Burrough	Technical requirements (hardware, software and data structures); cost; expertise; embedding GIS in the organization
1989	Aronoff	Technology; database creation; institutional barriers; expertise
1991	Atenucci *et al.*	Software (data structures, hypermedia, artificial intelligence); hardware (PC performance, mass storage); communications and networking; system implementation
1993	Van Oosterom	Software (data structure)
1995	Huxhold and Levinsohn	Embedding GIS technology in organizations; managing GIS projects
2001	Bregt *et al.*	Data availability and quality; model applicability and development

1.4   CONTENTS OF THE BOOK

The problem of how to classify models was posed by Crane and Goodchild (1993; p 481). In Chapter 2 a taxomony of GIS environmental models is proposed, and examples provided.

As identified in Table 1.2 and in Chapter 12, the availability of spatial data is a major bottleneck for managers wishing to implement GIS for environmental modelling. This important issue is discussed in Chapter 3 for remotely sensed data, and Chapter 4 for GIS data.

In Chapter 5, the application of GIS to global monitoring and environmental models is addressed. This chapter describes the problems and solutions for a number of pressing global issues. The scale changes in Chapter 6 from a global to a regional perspective, where environmental models for vegetation mapping and monitoring are discussed.

The emphasis in Chapter 7 shifts from vegetation models to habitat models in general, with special attention to wildlife habitat models.

The issue of biodiversity, and how geoinformation is used for developing policy about biodiversity, is addressed in Chapter 8.

An important branch of environmental modelling is hydrology; water quantity and quality is a primary determinant of productivity in agricultural as well as environmental systems. The use of GIS and remote sensing in hydrological modelling is discussed in Chapter 9.

Extreme weather events, as well as other natural hazards such as earthquakes, are described in Chapter 10. Techniques developed to model such hazards are presented.

The penultimate chapter (11) shows how geoinformation may be used in environmental impact assessment (EIS) and land use planning. A number of case studies are presented.

Finally in Chapter 12, the feedback from participants and lecturers on problems in the use of GIS and remote sensing for environmental modelling, and possible solutions to these problems is presented.

1.5 REFERENCES

Aronoff, S., 1989, *Geographic Information Systems: A Management Perspective*. Ottawa, WDL Publications

Antenucci, J. C., Brown, K., Croswell, P.L., Kevany, M.J., Archer, H. (1991). *Geographic information systems a guide to the technology*. New York, Chapman and Hall.

Barbier, E.B., 1987, The concept of sustainable development. *Environmental Conservation*, **14**: 101-110

Bregt, A., Skidmore, A.K. and G. Nieuwenhuis, 2001, *Environmental Modelling: Issues and Discussion*. Chapter 12 (this volume).

Burrough, P.A., 1986, *Principles of Geographic Information Systems for Land Resources Assessment*. Oxford, Clarendon Press.

Burrough, P. A. and McDonnell, R.A. (1993). *Principles of Geographical Information Systems*. Oxford, Clarendon Press

Crane, M.P. and Goodchild, M.F., 1993, Epilog In Goodchild, M.F., Parks, B.O., Steyart, L.T. (editors). *Environmental Modelling with GIS* (Oxford University Press, New York)

Duerr, W.A., Teeguarden, D.E. *et al.*, 1979, *Forest resource management: decision-making principles and case*. Philadelphia, W.B. Saunders

Fresco, L.O. and S.A. Kroonenberg, 1992, *Time and spatial scales in ecological sustainability*, pp. 155-168.

Garner, B., 1993, *Introduction to GIS*. UNSW Lecture notes, University of NSW, Sydney, NSW 2000.

M. Goodchild, B. P. a. L. S., Ed. (1993). *Environmental modeling with GIS*. Oxford, Oxford University Press.

Goodland, R. and G. Ledec, 1987, Neoclassical economics and principles of sustainable development. *Ecological Modelling*, **38**: 19-46

Heit, M. and A. Shortreid, 1991, *GIS Applications in Natural Resources*. GIS World: Fort Collins, Colorado

Huxhold, W.E. and A.G. Levinsohn, 1995, *Managing GIS Projects*. New York, Oxford University Press.

Jeffers, J.N.R., 1978, *An introduction to systems analysis: with ecological applications*. London, Edward Arnold.

Laurini, R. and D. Thompson, 1992, *Fundamentals of Spatial Information Systems*. London, Academic Press.

McCall, M., per comm. ITC, P.O. Box 6, 7500 AA Enschede, The Netherlands.

Odum, E.P., 1975, *Ecology*. London, Holt Reinhart and Winston.

Oosterom van, P. J. M. (1993). *Reactive data structures for geographic information systems*. Oxford, Oxford University Press.

Taxonomy of environmental models in the spatial sciences

Andrew K. Skidmore

2.1 INTRODUCTION

Environmental models simulate the functioning of environmental processes. The motivation behind developing an environmental model is often to explain complex behaviour in environmental systems, or improve understanding of a system. Environmental models may also be extrapolated through time in order to predict future environmental conditions, or to compare predicted behaviour to observed processes or phenomena. However, a model should not be used for both prediction and explanation tasks simultaneously.

Geographic information system (GIS) models may be varied in space, in time, or in the state variables. In order to develop and validate a model, one factor should be varied and all others held constant. Environmental models are being developed and used in a wide range of disciplines, at scales ranging from a few meters to the whole earth, as well as for purposes including management of resources, solving environmental problems and developing policies. GIS and remote sensing provide tools to extrapolate models in space, as well as to upscale models to smaller scales.

Aristotle wrote about a two-step process of firstly using one's imagination to inquire and discover, and a second step to demonstrate or prove the discovery Britannica 1989 14:67). This approach is the basis of the scientific approach, and is applied universally for environmental model development in GIS. In the section on empirical models, the statistical method of firstly exploring data sets in order to discover pattern, and then confirming the pattern by statistical inference, follows this process in a classical manner. But other model types also rely on this process of inquiry and then proof. For example, the section on process models shows how theoretical models based on experience (observation and/or field data) can be built.

Why spend time developing taxonomy of environmental models – does it serve any purpose except for academic curiosity? In the context of this book, taxonomy is a framework to clarify thought and organize material. This assists a user to easily identify similar environmental models that may be applied to a problem. In the same way, model developers may also utilise or adapt similar models. But taxonomy also gives an insight to very different models, and hopefully helps in transferring knowledge between different application areas of the environmental sciences.

2.2 TAXONOMY OF MODELS

Using terminology found in the GIS and environmental literature, models are here characterized as 'models of logic' (inductive and deductive), and 'models based on processing method' (deterministic and stochastic) (see Table 2.1). The deterministic category has been further subdivided into empirical, process and knowledge based models (Table 2.1). The sections of this chapter describe the individual model type; that is, a section devoted to each column of Table 2.1 (e.g. see 2.3.2 for inductive models) or row (e.g. see 2.4.1 for deterministic-empirical models). In addition, an example of an environmental application is cited for each model.

An important observation from Table 2.1 is that an environmental model is categorized by both a processing method and a logic type. For example, the CART model (see 2.3.2) is both deterministic (empirical) as well as inductive. In categorizing models based on this taxonomy, it is necessary to cite both the logic model and the processing method.

Finally, a model may actually be a concatenation of two (or more) categories in Table 2.1.

Table 2.1: A taxonomy of models used in environmental science and GIS.

			Model of logic (see Section 2. 3)	
			Deductive (see Section 2.3.1)	**Inductive (see Section 2.3.2)**
Model based on processing method (see Section 2.4)	**Deterministic (see Section 2.4)**	**Empirical (see Section 2.4.1)**	Modified inductive models (e.g. R-USLE); process models classification by supervised classifiers (model inversion)	Statistical models (e.g. regression such as USLE); training of supervised classifiers (e.g. maximum likelihood) threshold models (e.g. BIOCLIM) rule induction (e.g. CART) Others: geostatistical models, Genetic algorithms
		Knowledge (see Section 2.4.2)	Expert system (based on knowledge generated from experience)	Bayesian expert system; fuzzy systems
		Process (see Section 2.4.3)	Hydrological models Ecological models	Modification of inductive model coefficients for local conditions by use of field or lab data
	Stochastic (see Section 2.5)		Monte Carlo simulation	Neural network classification; Monte Carlo simulation

For example, a model may be a combination of an inductive-empirical and a deductive-knowledge method. Care must be taken to identify the components of the model, otherwise the taxonomic system will not work. This point is addressed further in the chapter.

2.3 MODELS OF LOGIC

2.3.1 Deductive models

A deductive model draws a specific conclusion (that is generates a new proposition) from a set of general propositions (the premises). In other words, deductive reasoning proceeds from general truths or reasons (where the premises are self-evident) to a conclusion. The assumption is that the conclusion necessarily follows the premises; that is, if you accept the premises, then it would be self-contradictory to reject the conclusion.

An example of deduction is the famous Euclid's 'Elements', a book written about 300 BC. Euclid first defines fundamental properties and concepts, such as point, line, plane and angle. For example, a line is a length joining two points. He then defines primitive propositions or postulates about these fundamental concepts, which the reader is asked to consider as true, based on their knowledge of the physical world. Finally, the primitive propositions are used to prove theorems, such as Pythagoras' theorem that the sum of the squares of a right-angled triangle equals the square of the length of the hypotenuse. In this manner, the truth of the theorem is proven based on the acceptance of the postulates.

Another example of deduction is the modelling of feedback between vegetation cover, grazing intensity and effective rainfall and development of patches in grazing areas (Rietkerk 1998). In Figure 2.1 (taken from Rietkerk *et al.* 1996 and Rietkerk 1998), the controlling variables are rainfall and grazing intensity, while the state variable is the vegetation community. State I in Figure 2.1 are perennial grasses, state II are annual grasses and state III are perennial herbs. The diagram links together a number of assumptions and propositions (taken from the literature) about how a change in rainfall and grazing intensity will alter the mix of the state variables (*viz.* perennial grass, annual grass, and perennial herbs). For example, it is assumed that the three vegetation states are system equilibria. Rietkerk *et al.* (1996) show that according to the literature this is a reasonable assumption; the primeval vegetation of the Sahel at low grazing intensities is a perennial grass steppe. They go on to discuss the various transition phases between the three vegetation states and to support their conclusion that Figure 2.1 is reasonable they cite propositions from the literature.

For example, transition 'T2a' in Figure 2.1, is a catastrophic transition where low rainfall is combined with high grazing, leading to rapid transition of perennial grass to perennial herbs, without passing through the annual grass stage II. Such deductive models have been rarely extrapolated in space.

In all these examples, the deductive model is based on plausible physical laws. The mechanism involved in the model is also described.

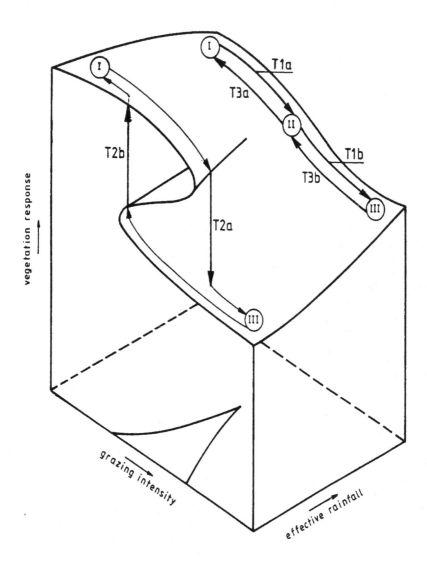

Figure 2.1: The cusp catastrophe model applied to the Sahelian rangeland dynamics (from Rietkerk *et al.* 1998).

2.3.2 Inductive models

The logic of inductive arguments is considered synonymous with the methods of natural, physical and social sciences. Inductive arguments derive a conclusion from particular facts that appear to serve as evidence for the conclusion. In other words, a series of facts may be used to derive or prove a general statement. This implies that based on experience (usually generated from field data), induction can lead to the discovery of patterns. The relationship between the facts and the conclusion is observed, but the exact mechanism may not be understood. For example, it may be found from field observation or sampling that a tree (*Eucalyptus sieberi*) frequently occurs on ridges, but such an observation does not explain the occurrence of this species at this particular ecological location.

As noted above, induction is considered to be an integral part of the scientific method and typically follows a number of steps:

- Defining the problem using imagination and discovery.
- Defining the research question to be tested.
- Based on the research question, defining the research hypotheses that are to be proven.
- Collecting facts, usually by sampling data for statistical testing.
- Exploratory data analysis, whereby patterns in the data are visualized.
- Confirmatory analysis rejects (or fails to reject) the research hypothesis at a specified level of confidence and draws a conclusion.

The inductive method as adopted in science, and formalized in statistics, claim that the use of facts (data) leads to an ability to state a probability (that is a confidence or level of reasonableness) about the conclusion.

An example of an inductive model is the classification and regression tree (CART) method also known as a decision tree (Brieman *et al.* 1984; Kettle 1993; Skidmore *et al.* 1996). It is a technique for developing rules by recursively splitting the learning sample into binary subsets in order to create the most homogenous (best) descendent subset as well as a node (rule) in the decision tree (Figure 2.2a) (see Brieman *et al.* 1984; and Quinlan 1986 for details about this process). The process is repeated for each descendent subset, until all objects[1] are allocated to a homogenous subset. Decision rules generated from the descending subset paths are summarized so that an unknown grid cell may be passed down the decision tree to obtain its modelled class membership (Quinlan 1986) (Figure 2.2b). Note that in Figure 2.2a, the distribution of two hypothetical species ($y = 0$ and $y = 1$) is shown with gradient and topographic position, where topographic position 0 is a ridge, topographic position 5 is a gully, and values in between are midslopes. The data set is split at values of gradient $= 10°$ and topographic position $= 1$.

The final form of the decision tree is similar to a taxonomic tree (Moore *et al.* 1990) where the answer to a question in a higher level determines the next question asked. At the leaf (or node) of the tree, the class is identified.

---

[1] For example in the paper by Skidmore *et al.* (1996), the objects were kangaroos.

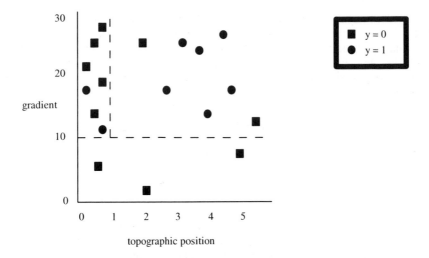

Figure 2.2a: The distribution of two hypothetical species (y = 0 and y = 1) is shown with gradient and topographic position, where topographic position 0 is a ridge, topographic position 5 is a gully, and values in between are midslopes.

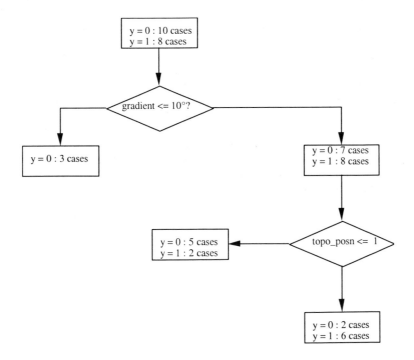

Figure 2.2b: The decision tree rules generated from the data distribution in Figure 2.2a.

2.3.3 Discussion

Both inductive and deductive methods have been used for environmental modelling. However, inductive models dominate spatial data handling (GIS and remote sensing) in the environmental sciences. As stated in 2.2, some models are a mix of methods; a good example of a mix of inductive and deductive methods is a global climate model (see also Chapter 4 by Reed *et al.* as well as Chapter 5 by Los *et al.*). In these models, complex interactions within and between the atmosphere and biosphere are described and linked. For example, photosynthesis is calculated as a function of absorbed photosynthetically active radiation (APAR), temperature, day length and canopy conductance of radiation. A component of this calculation is the daily net photosynthesis, the rationale for which is given by Hazeltine (1996). Some of the parameters in this calculation of daily net photosynthesis may be estimated from remotely sensed data (such as the fraction of photosynthetically active radiation) or interpolated from weather records (such as daily rainfall), while other constants are estimated from laboratory experiments (e.g. a scaling factor for the photosynthetic efficiency of different vegetation types). Thus the formula has been deduced, but the components of the formulae that include constants and variable coefficients are calculated using induction.

Classification problems may be considered to be a mix of deductive and inductive methods. The first stage of a classification process is inductive, where independent data (usually collected in the field or obtained from remotely sensed imagery) are explored for possible relationships with the dependent variable(s) that is to be modelled. For example, if land cover is to be classified from satellite images, input data are collected from known areas and used to estimate parameters of a particular image classifier algorithm such as the maximum likelihood classifier (Richards 1986). The second stage of the supervised classification process is deductive. The decision rules (premises) generated in the first phase are used to classify an unknown pixel element, and come up with a new proposition that the pixel element is a particular ground cover. Thus, the classification of remotely sensed data is in reality two (empirical) phases – the first phase (training) uses induction and the second phase (classification) uses deduction.

Another example of a combined inductive-deductive model in GIS may be based on a series of rules (propositions) that a GIS analyst believes are important in determining a process or conclusion. For example, a model has been developed to map the dominant plant type at a global scale (Hazeltine 1996). The model is deduced from propositions linking particular biome types (e.g. dry savannas) to a number of independent variables including:

- leaf area index
- net primary production
- average available soil moisture
- temperature of the coldest month
- mean daily temperature
- number of days of minimum temperature for growth.

The thresholds for the independent variable determining the distribution of the biome type are induced from observations and measurements by other ecologists.

For example, dry savannas are delineated by a leaf area index of between 0.6 and 1.5, and by a monthly average available soil moisture of greater than 65%.

A well-known philosophy in science, developed by Popper, rejects the inductive method for the physical (environmental) sciences and instead advocates a deductive process in which hypotheses are tested by the 'falsifiability criterion'. A scientist seeks to identify an instance that contradicts a hypothesis or postulated rule; this observation then invalidates the hypothesis. Putting it another way, a theory is accepted if no evidence is produced to show it is false.

2.4 DETERMINISTIC MODELS

A deterministic model has a fixed output for a specific input. Most deterministic models are derived empirically from field plot measurements, though rules or knowledge may be encapsulated in an expert system and will consistently generate a given output for a specific input. Deterministic models may be inductive or deductive.

2.4.1 Empirical models

Empirical models are also known as statistical, numerical or data driven models. This type of model is derived from data, and in science the model is usually developed using statistical tools (for example, regression). In other words, empiricism is that beliefs may only be accepted once they have been confirmed by actual experience. As a consequence, empirical models are usually site-specific, because the data are collected 'locally'. The location at which the model is developed may be different to other locations (for example, the climate or soil conditions may vary), so empirical models of the natural environment are not often applicable when extrapolated to new areas.

For empirical models used in the spatial sciences, models are calculated from (training) data collected in the field. Recall that inductive models also use training data, so a model may be classified as inductive-empirical (see 2.3.2). However, not all inductive models are empirical (see Table 2.1)!

Statistical tests (usually employed to derive information and conclusions from a database) require a proper sampling design, for example that sufficient data be collected, as well as certain assumptions be met such as data are drawn independently from a population (Cochran 1977). A variety of statistical methods have been used in empirical studies, and some authors have proposed that empirical models be subdivided on the basis of statistical method. Burrough (1989) distinguished between regression and threshold empirical models; these are two dominant techniques in GIS. An example of a regression model is the Universal Soil Loss Equation (USLE), which was developed empirically using plot data in the United States of America (Hutacharoen 1987; Moussa, *et al.* 1990). In contrast, threshold models use boundary values to define decision surfaces and are often expressed using Boolean algebra. For example, dry savannas in the global vegetation biome map cited in 2.3.3 (Hazeltine 1996) are defined using a number of factors including the leaf area index of between 0.6 and 1.5. Other examples of empirical models where thresholds are used include CART (see 2.3.2) and BIOCLIM.

The BIOCLIM system (see also Chapter 8 by Busby) determines the distribution of both plants and animals based on climatic surfaces. Busby (1986) predicted the distribution of *Nothofagus cunninghamiana* (Antarctic Beech), the Long-footed Potoroo (*Potorous longipes*), and the Antilopine Wallaroo (*Macropus antilopinus*), and inferred changes to the distribution of these species in response to change in mean annual temperature resulting from the 'greenhouse effect'. Nix (1986) mapped the range of elapid snakes. Booth *et al.* (1988) used BIOCLIM to identify potential *Acacia* species suitable for fuel-wood plantations in Africa, and Mackay *et al.* (1989) classified areas for World Heritage Listing. Skidmore *et al.* (1997) used BIOCLIM to predict the distribution of kangaroos.

The basis of BIOCLIM is the interpolation of climate variables over a regular geographical grid. If a species is sampled over this grid, it is possible to model the species response to the interpolated climate variables. In other words, the (independent) climate variables determine the (dependent) species distribution. The climate variables used in BIOCLIM form an environmental envelope for the species. Firstly, the BIOCLIM process involves ordering each variable. Secondly, if the climate value for a grid cell falls within a user-defined range (for example, the 5th and 95th percentile) for each of the climatic variables being considered, the cell is considered to have a suitable climate for the species. Using a similar argument, if the cell values for one (or more) climatic variables fall outside the 95th percentile range but within the (minimum) 0-5th percentile and (maximum) 95-100th percentile, the cell is considered marginal for a species. Cells with values falling outside the range of the sampled data (for any of the climatic variables) are considered unsuitable for the species (Figure 2.3).

In practice, there are other types of empirical models, including genetic algorithms (Dibble and Densham 1993) and geostatistical models (Varekamp *et al.* 1996). These, and other, models do not fit into the regression or threshold categories for inductive and empirical models as proposed by Burrough (1989), so it is considered simpler and more robust not to subdivide empirical models further.

Bonham-Carter (1994) grouped empirical and inductive models into two types, *viz.*, exploratory and confirmatory. This follows the established procedure in statistics of using exploratory data analysis (EDA) followed by confirmatory methods (Tukey 1977). In exploratory data analysis, data are examined in order that patterns are revealed to the analyst. Graphical methods are usually employed to visualize patterns in the data (for example, box plots or histograms). Most modern statistical packages permit a hopper-feed approach to developing insights about relationships in the data.

In other words, all available data are fed in the system, data are explored, and it is hoped that something meaningful emerges.[2] Once relationships are discovered, data driven empirical methods usually confirm rules, processes or relationships by statistical analysis.

---

[2] An approach frowned upon by some scientists who believe that science should be driven by questions and hypotheses that determine which data are collected, and pre-define the statistical methods used to confirm relationships within the data set.

An example is taken from Ahlcrona (1988) who identified a linear relationship between the normalized difference vegetation index[3] (NDVI) calculated using Landsat MSS (multispectral scanner) imagery and wet grass biomass (Figure 2.4).

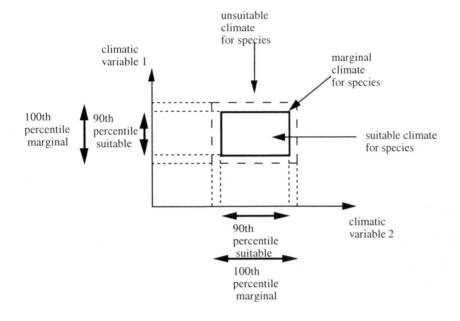

Figure 2.3: Possible BIOCLIM class boundaries for two climatic variables.

Regression was used to calculate a linear model between the dependent (wet grass biomass) and independent (MSS NDVI) variables with a correlation coefficient of 0.61.

A derivative of the Universal Soil Loss Equation (USLE) is the Revised Universal Soil Loss Equation (RUSLE), which is used to calculate sheet and rill erosion (Flacke *et al.* 1990; Rosewell *et al.* 1991). The RUSLE model is an interesting example of a localized empirical model that has been modified (using deduction) and then reapplied in new locations.

---

[3] NDVI is a deduced relationship between the infrared and red reflectance of objects or land cover.

$$NDVI = \frac{NIR - red}{NIR + red}$$

where NIR is the reflectance in the near infrared channel and red is the reflectance in the red channel.

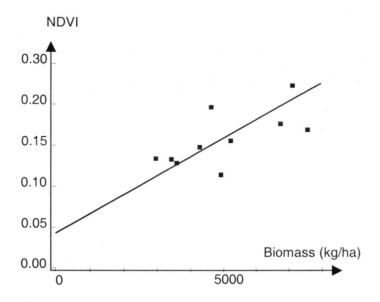

Figure 2.4: The relationship between MSS NDVI and wet grass biomass (from Ahlcrona 1988).

2.4.2 Knowledge driven models

Knowledge driven models use rules to encapsulate relationships between dependent and independent variables in the environment. Rules can be generated from expert opinion, or alternatively from data using statistical induction (such as CART described in 2.3.2). The rules can directly classify (unknown) spatial objects (grid cells or polygons) by deduction, or the rules may be input to an expert system. An expert system is a type of knowledge driven model.

An expert system comprises a knowledge base of rules, a method for processing the rules (the inference engine), an interface to the user, and the (independent) spatial data that are usually stored in a GIS. The structure of the knowledge base largely determines the appropriate inference technique required to generate a conclusion from the expert system. One common method for representing knowledge is the frame (Forsyth 1984), while a method called a probability matrix has also been developed (Skidmore 1989).

The advantage of the frame structure is that knowledge is organized around objects, and knowledge may be inherited from one frame to the next. This is similar to our own 'memory', where knowledge or facts are often remembered through association with other knowledge. The frame structure has been utilized in some expert system applications (Skidmore *et al.* 1992). A second method of representing knowledge in a GIS, called a probability matrix, links the probability of a species occurring at different environmental positions (Skidmore 1989).

Expert systems have been developed from, and given a theoretical foundation based on the field of, formal logic. Following the definitions given in the 'inductive logic' section above (see 2.3.2), formal logic is used to infer a conclusion from facts contained within the knowledge base. For example, given the evidence that a location is a ridge top, and given that if there is a ridge then *Eucalyptus sieberi* occurs, it is possible to infer (conclude) that *Eucalyptus sieberi* is present on the ridge. Using this flow of logic (*modus tollens*), the evidence (E) that a ridge occurs may be linked with a hypothesis (H) that *Eucalyptus sieberi* is present, using an expert system. In expert systems, the evidence (E) is often called an antecedent, and the hypothesis (H) the consequent. In other words, given evidence (E) occurs then conclude the hypothesis (H):

$$\text{GIVEN} \quad \rightarrow \quad \text{E} \quad \rightarrow \quad \text{THEN} \quad \rightarrow \quad \text{H}$$

$$\qquad\qquad\qquad \text{antecedent} \qquad\qquad\qquad\quad \text{consequent}$$

$$\qquad\qquad\qquad \text{evidence} \qquad\qquad\qquad\qquad \text{hypothesis}$$

where E is the evidence, H is the hypothesis.

Two methods exist for linking the evidence with the hypotheses. The first is forward chaining, where the inference works forward from the evidence (e.g. data represented at a grid cell) to the hypothesis. This is a 'data driven' process, where given some evidence, a hypothesis is inferred from the expert's rules and is an inductive model. The second method is simply the reverse, and is called backwards chaining. In other words, given a hypothesis, the expert system examines how much evidence there is to support the hypothesis. Backwards chaining is obviously a hypothesis driven process, and is akin to the deductive model as described in 2.3.1. But what happens when you do not know with 100 per cent confidence whether the rules are true? For example, *Eucalyptus sieberi* may be present only on some ridges in an area of interest. In such a case you need a method to handle uncertainty in the rules, so that the rules may be weighted on the basis of the uncertainty.

The basis of the Bayes' inferencing algorithm is that knowledge about the likelihood of a hypothesis occurring, given a piece of evidence, may be thought of as a conditional probability. For example, a user may not be certain whether *Eucalyptus sieberi* always occurs on ridges — it may sometimes occur on midslopes. This knowledge may be expressed as the user being reasonably certain (e.g. a weight of 0.9) that *Eucalyptus sieberi* occurs on ridges. By linking the knowledge (weights) with GIS layers, the attributes of the raster cell or polygon are matched with the information in the knowledge (rule) base. The expert system then infers the most likely class at a given cell, using Bayes' Theory.

The expert system was executed and a soil type map predicted by an expert system was plotted for a catchment in south eastern Australia (Skidmore *et al.* 1996). When compared with a soil type map of the same soil classes as prepared by a soil scientist, it was obvious that the two results are similar. 53 soil pits were dug through the area, and 73.6 per cent of the pits were correctly predicted by the expert system. There was no statistically significant difference between the accuracy of the expert system map and the map prepared by the soil scientist, as tested by the Kappa statistic (Cohen 1960).

The Bayesian expert system described above is inductive, as input data from field plots are used to develop rules. It is also possible to develop rules for an expert system based only on existing knowledge; that is an expert would deduce a model about an environmental system. Such an expert system is deterministic, knowledge based, and of course deductive (see Table 2.1). As noted in 2.2, environmental models may be a mix of categories (Table 2.1).

2.4.3   Process driven models

Process driven models, also known as conceptual models, physically based models, process driven systems, white box models (as opposed to 'black box' because the process is understood) or goal driven systems, use mathematics (often supported by graphical examples) to describe the factors controlling a process. Process driven models are mostly deductive, and to a large extent the features of deductive models described in 2.3.1 are applicable. This class of models describe a process based on understanding and established concepts (prepositions), though parameter values may be estimated from data. In many respects, a process model is a pure science product. However, induction is also frequently used to support the development of process driven models particularly to estimate the value of the model parameters, or to refine the underlying concepts (or factors) on which the model is constructed. The necessity to input detailed parameters that are frequently not available make the task of operating and validating process-models difficult. In practice, most process models are limited to small, relatively simple areas (Pickup and Chewings 1986; Pickup and Chewings 1990; Moore *et al.* 1993; Riekerk *et al.* 1998)

Process models may be static or dynamic with respect to time. Static process models split complex areas of land into relatively homogeneous sub-units, and then use the output from one sub-unit as an input to the next sub-unit (e.g. O'Loughlin 1986). Dynamic process models iterate the process over time and typically attempt to represent a continuous surface.

An example of a process model based on deduction is the Hortonian overland flow model (Horton 1945):

$$Q = (I-F)A \quad (2)$$

Where Q is the surface runoff rate, I is the rainfall intensity and F represents the infiltration rate and A is the catchment area. The generality of Hortonian overland flow has been criticised because:

- surface runoff is dependent on ground conditions, which vary spatially and over time
- that the calculation of surface runoff from comparisons of rainfall intensity and infiltration rates holds good only for very small areas
- that the Hortonian overland flow assumes average conditions over an entire catchment
- the independent parameters (i.e., I, F and A) in equation 2 require induction to estimate their coefficients.

Hortonian overland flow is an example of a lumped empirical model, where the output is calculated for a region based on average input values for the region and is akin in GIS to polygon data structures.

In contrast to lumped models, distributed process models assume that space is continuous, and calculations are made for each element within the area. The elements may be linked in order to estimate the movement between elements (for example, the flow of water between elements in a hydrological model, or the movement of air in a global climate model). Distributed models are developed using raster GIS. The technology makes it simple to spatially and temporally link elements, allowing models to describe the flow of materials or water over a landscape. Such grid based models have been widely developed in hydrology (e.g. TOPMODEL, SHE, ANSWERS).

The problem with distributed models is that they frequently require a large number of input variables of a specific resolution. Remote sensing data, or geostatistics, therefore generate these spatially distributed variables. However, major obstacles exist to the use of distributed models including:

- scaling up (e.g. from points to catchments to continents)
- models based on point data may not be applicable
- input data vary in scale and accuracy (garbage in — garbage out).

As a number of researchers have noted, there is little evidence that complex process models are superior to simple empirical models for many environmental modelling applications (Burrough *et al.* 1996).

Based on the evidence presented in 2.3.1 and 2.4.3, it would be tempting to simplify the taxonomy system and merge 2.4.3 into 2.3.1 (Table 2.1). However, the widespread use of the term 'process driven model' in hydrology, and the fact that process driven models is a hybrid consisting of a concatenation of a number of models (see 2.2), on balance resulted in this category of model remaining separate.

2.5 STOCHASTIC MODELS

If the input data, or parameters of the model itself, are (randomly) varied then the output also varies. A variable output is the essence of a stochastic model.

An example of a stochastic model increasingly used in environmental modelling is the neural network model, commonly implemented using the back-propagation (BP) algorithm. The structure of a typical three-layered neural network is shown in Figure 2.5; however networks may easily be constructed with more than three layers.

To train a network, a grid cell is presented with values derived from a GIS. For example, in Figure 2.5, the values for a cell may be elevation equal to 0.8, aspect equal to 0.3 and SPOT visible band equal to 0.5 (note the input values are normalized to range between 0 and 1). Simultaneously, an output class is presented to the network; the output node has an associated output, or target, value. In other words, an output class, such as water, may be assigned to an output node number (for example node 3 in Figure 2.5), and given a target value of, for example, 0.90. Clearly, the neural network is trained using induction (see 2.3.2).

The BP algorithm iterates in a forward and then in a backward direction. During the forward step, the values of the output nodes are calculated from the input layer. Phase two compares the calculated output node values to the target (i.e. known) values. The difference is treated as error, and this error modifies connection weights in the previous layer. This represents one epoch of the BP algorithm. In an iterative process, the output node values are again calculated, and the error is propagated backwards. The BP algorithm continues until the total error in the system decreases to a pre-specified level, or the rate of decrease in the total system error becomes asymptotic. Prior to the first epoch, the neural network algorithm assigns random weights to the nodes and introduces the stochastic element to the neural network model.

Node weights are an interesting neural network parameter to adjust (Skidmore *et al.* 1997). An experimental set up was chosen that produced an accurate map of forest soil, and the network parameters were noted (Skidmore *et al.* 1997).

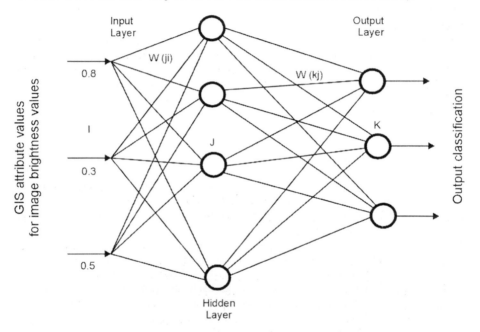

Figure 2.5: Neural network structure for the BP algorithm.

A map of the classes predicted by the neural network shows the classification was reasonable. All network parameters were then held constant (e.g. number of learning patterns, number of nodes, number of layers, learning rate, momentum etc.), except that the starting weights were randomly adjusted by ± 5%. Five different maps were produced, with each map having slightly different starting weights. Even though the accuracy of the training and test data is similar (ranging from 90 to 97 per cent training accuracy and 42 to 55 per cent test accuracy), the spatial distribution of the classes was quite different. Such a variation in mapping accuracy highlights the stochastic nature of neural networks.

Stochastic models have also been developed where the average (and variance) value for many (usually random) events are calculated. For example, randomly selecting the input data from a known population distribution, and then noting the range of output values obtained, indicates the possible range of output values, as well as the distribution of the output.

2.6 CONCLUSION

A taxonomy of GIS models has been presented with examples from various application fields in the environmental sciences. Some of the model types have had limited application in the spatial sciences. Other model types are widely applied, such as inductive empirical models.

As highlighted in this chapter, many environmental applications combine two (or more) categories (as detailed in Table 2.1), though the modelling process may appear seamless to a user. In order to use the taxonomic system, a user must deconstruct the application, and identify the taxonomic categories. This provides the user with a framework to clarify thoughts and organize material. In other words, a user, or model developer, can easily identify similar environmental models that may be applied or adapted to a problem. As taxonomy also gives an insight to very different models, a taxonomy hopefully helps in transferring knowledge between different application areas of the environmental sciences.

2.7 REFERENCES

Ahlcrona, E., 1988, *The impact of climate and man on land transformation in central Sudan.* PhD thesis. Lund University, Lund, Sweden.

Aleksander, I. and Morton, H., 1990, *An introduction to neural computing.* London, Chapman and Hall.

Bonham-Carter, G.F., 1994, Methods of spatial data integration for mineral mapping potential. *Proceedings Seventh Australasian Remote Sensing Conference,* Melbourne, Aust. Soc. for Remote Sensing and Photogrammetry.

Booth, T.H., Nix, H.A., Hutchinson, M.F., and Jovanovic, T., 1988. Niche analysis and tree species introduction. *Forest Ecology and Management* 23:47–59

Brieman, L., Friedman, J.H., Ollshen, R.A., and Stone, C.J., 1984, *Classification and Regression Trees.* Belmont, CA, Wadsworth.

Britannica, E., 1989, *The New Encyclopaedia Britanica.* Chicago, Encyclopeadia Britannica.

Burrough, P.A., 1989, Matching spatial database and quantitative models in land resource assessment. *Soil Use and Management* 5: 3–8

Busby, J.R., 1986. A bioclimatic analysis of *Nothofagus cunninghamia* in south eastern Australia. *Australian Journal of Ecology* 11:1–7

Burrough, P.A., van Rijn, R. *et al.*, 1996, Spatial data quality and error analysis issues: GIS functions and environmental modelling. In: Goodchild, M.F., Steyaert, L.T., Parks B.O., (ed.). *GIS and Environmental Modelling: Progress and Research Issues.* Fort Collins, CO, GIS World Books, 29–34.

Cochran, W.G., 1977, *Sampling Techniques.* New York, Wiley.

Dibble, C. and Densham, P., 1993, Generating interesting Alternatives in GIS and SDSS Using Genetic Algorithms. In *Proceedings of the GIS/LIS '93 Conference*, Minneapolis, ACSM-ASPRS-URISA-AM/FM.

Flacke, W., Auerswald, K. *et al.*, 1990, Combining a Modified Universal Soil Loss Equation with a Digital from Rain Wash. *Catena*, **17**: 383–397.

Forsyth, R., 1984, *Expert Systems: Principles and Case Studies*. London, Chapman and Hall.

Hazeltine, A., 1996, *Modelling the vegetation of the Earth*. PhD Thesis. Lund University, Lund, Sweden.

Horton, R.E., 1945, Erosional development of streams and their drainage basins: hydrophysical approach to quantitative morphology. *Bulletin of the Geological Society of America*, **56**: 275–370.

Hutacharoen, M., 1987, Application of Geographic Information Systems Technology to the Analysis of Deforestation and Associated Environmental Hazards in Northern Thailand. In *Proceedings GIS '87, San Francisco, California*. Washington, American Society of Photogrammetry and Remote Sensing.

Kettle, S., 1993, CART learns faster than knowledge seeker. In *Proceedings of the Conference on Land Information Management, Geographic Information Systems and Advanced Remote Sensing* **2**: 161–171. School of Surveying, UNSW, P.O. Box 1, Kensington, NSW 2033, Australia.

Kosko, B., 1992, *Neural networks and fuzzy systems: a dynamical systems approach to machine intelligence*. Englewood Cliffs, New Jersey, Prentice Hall.

Kramer, M.R. and Lembo, A.J., 1989, *A GIS Design Methodology for Evaluating Large-Scale Commercial Development*. In *Proceedings GIS/LIS '89*, Orlando, Florida, ASPRS, AAG, URPIS, AM/FM International.

Mackay, B.G., Nix, H.A., Stein, J.A., Cork, S.E., and Bullen, F.T., 1989. Assessing the representatives of the wet tropics of north Queensland world heritage property. *Biological Conservation* **50**:279–303

Maidment, D.R., 1993, GIS and hydrologic modelling. In: Goodchild, M.F., Parks, B.O., and Steyaert, L.T (eds.). *Environmental modelling with GIS*. Oxford Oxford University Press.

Moore, D.M., Lees, B.G. and Davey, S.M., 1990, A New Method for predicting vegetation distributions using decision tree analysis in a GIS. *Environmental Management*, **15**: 59–71.

Moore, D.M, Lees, B.G., *et al.*, 1991, A New Method for Predicting Vegetation System. *Environmental Management*, **15**: 59–71.

Moore, I.D., Turner, A.K. *et al.*, 1993, GIS and land-surface-subsurface process modelling. In: Goodchild, M.F., Parks, B.O. and Steyaert, L.T. (ed.). *Environmental modelling with GIS*. Oxford, Oxford University Press.

Moussa, O.M., Smith, S.E. *et al.*, 1990, GIS for Monitoring Sediment-Yield from Large Watershed. In *Proceedings of GIS/LIS '90*, Anaheim California, ASP&RS, AAG, URPIS and AM/FM International.

Nix, H., 1986. A biogeographical analysis of Australian elapid snakes. In: Longmore, R., (ed.), *Atlas of Elapid Snakes of Australia*, Bureau of Flora and Fauna, Canberra, Australian Government Publishing Service, 4–15.

O'Loughlin, E.M., 1986, Prediction of surface saturation zones in natural catchments by topographic analysis. *Water Resources Research*, **22**: 794 – 804.

Pao, Y.H., 1989, *Adaptive pattern recognition and neural networks*. Reading, Addison-Wesley.

Pickup, G. and Chewings, V.H., 1986, Random field modelling of spatial variations in erosion and deposition in flat alluvial landscapes in arid central Australia. *Ecol Model*, **33**: 269–69.

Pickup, G. and Chewings, V.H., 1990, Mapping and Forecasting Soil Erosion Patterns from Landsat on a Microcomputer-Based Image Processing Facility. *Australian Rangeland Journal*, **8**: 57–62.

Quinlan, J.R., 1986, Introduction to decision trees. *Machine Learning*, **1**: 81–106.

Richards, J.A., 1986, *Remote Sensing – digital analysis*. Berlin, Springer-Verlag.

Rietkerk, M., 1998, *Catastrophic vegetation dynamics and soil degradation in semi-arid grazing systems*. PhD Thesis. Wageningen Agricultural University, Wageningen, The Netherlands.

Rietkerk, M., Ketner, P., Stroosnijnder, L., and Prins, H.H.T., 1996, Sahelian rangeland development: a catastrophe? *Journal of Range Management*, **49**: 512–519.

Rosewell, C.J., Crouch, R.J. *et al.*, 1991, Forms of erosion. In: Charman, P.E.V. and Murphy, B.W. *Soils – their properties and management: A soil conservation handbook for New South Wales*. Sydney, Sydney University Press.

Skidmore, A.K., 1989, An expert system classifies eucalypt forest types using Landsat Thematic Mapper data and a digital terrain model. *Photogrammetric Engineering and Remote Sensing*, **55**: 1449–1464.

Skidmore, A.K., 1990, Terrain Position as Mapped from a Gridded Digital Elevation Model. *International Journal of Geographical Information Systems*, **4**: 33–49.

Skidmore, A.K., Baang, J. and Luchananurug, P., 1992, Knowledge based methods in remote sensing and GIS. *Proceedings Sixth Australasian Remote Sensing Conference*, Wellington, New Zealand, **2**: 394–403.

Skidmore, A.K., Ryan, P.J., Short, D. and Dawes, W., 1991, Forest soil type mapping using an expert system with Landsat Thematic Mapper data and a digital terrain model. *International Journal of Geographical Information Systems*, **5**: 431–445.

Skidmore, A.K., Gauld, A., and Walker P.A., 1996, A comparison of GIS predictive models for mapping kangaroo habitat. *International Journal of Geographical Information Systems*, **10**: 441–454.

Skidmore, A.K., Turner, B.J., Brinkhof, W. and Knowles, E., 1997, Performance of a neural network mapping forests using GIS and remotely sensed data. *Photogrammetric Engineering and Remote Sensing*, **63**: 501–514.

Tukey, J., 1977, *Exploratory data analysis*. Reading, Addison-Wesley.

Varekamp, C., Skidmore, A.K. *et al.*, 1996, Using public domain geostatistical and GIS software for spatial interpolation. *Photogrammetric Engineering and Remote Sensing*, **62**: 845–854.

Walker, P.A., and Moore, D.M., 1988, SIMPLE: An inductive modelling and mapping tool for spatially-oriented data. *International Journal of Geographical Information Systems*, **2**: 347–364.

New environmental remote sensing systems

F. van der Meer, K.S. Schmidt, W. Bakker and W. Bijker

3.1 INTRODUCTION

Remote sensing can be defined as the acquisition of physical data of an object with a sensor that has no direct contact with the object itself. Photography of the Earth's surface dates back to the early 1800s, when in 1839 Louis Daguerre publicly reported results of images from photographic experiments. In 1858 the first aerial view from a balloon was produced and in 1910 Wilber Smith piloted the plane that acquired motion pictures of Centocelli in Italy. Image photography was collected on a routine basis during both world wars; during World War II non-visible parts of the electromagnetic (EM) spectrum were used for the first time and radar technology was introduced. In 1960s, the first meteorological satellite was launched, but actual image acquisition from space dates back to earlier times with various spy satellites. In 1972, with the launch of the earth observation land satellite Landsat 1 (renamed from ERTS-1), repetitive and systematic observations were acquired. Many dedicated earth observation missions followed Landsat 1 and in 1980 NASA started the development of high spectral resolution instruments (hyperspectral remote sensing) covering the visible and shortwave infrared portions of the EM spectrum, with narrow bands allowing spectra of pixels to be imaged (Goetz *et al.* 1985). Simultaneously in the field of active microwave remote sensing, research led to the development of multi-polarization radar systems and interferometric systems (Massonnet *et al.* 1994). The turn of the millennium marks the onset of a new era in remote sensing when many experimental sensors and system approaches will be mounted on satellites, thereby providing ready access to data on a global scale. Interferometric systems will provide global digital elevation models, while spaceborne hyperspectral systems will allow detailed spectrophysical measurements at almost any part of the earth's surface.

This chapter provides an overview of existing and planned satellite-based systems subdivided into the categories of high spatial resolution systems, high spectral resolution systems, high temporal resolution systems and radar systems (Figure 3.1). More technical details of some of these systems can be found in Kramer (1996). For readers requiring details of existing remote sensing systems as well as historical image archives, please refer to the references and internet links provided at the end of the chapter. The different sensor systems are catalogued within the internet links provided according to the order in which they are treated in the text. A brief discussion on the various application fields for the sensor types will follow the technical description of the instruments. The chapter provides a few classical references that serve as a starting point for further studies without attempting to be complete. In addition, cross references to other chapters in this book serve as a basis for a better understanding of the diversity of applications.

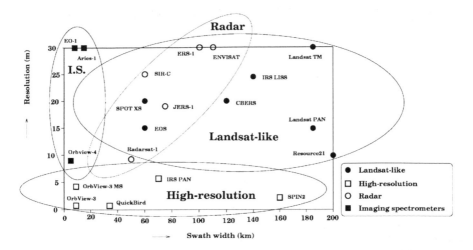

Figure 3.1: Classification of sensors.

3.2 HIGH SPATIAL RESOLUTION SENSORS

3.2.1 Historical overview

High spatial resolution sensors have a resolution of less than 5 m and were once the exclusive domain of spy satellites. In the 1960s, spy satellites existed that had a resolution better than 10 meters. Civil satellites had to wait until the very last days of the 20th century. The major breakthrough was one of policy rather than technology. The US Land Remote Sensing Act of 1992 concluded that a robust commercial satellite remote-sensing industry was important to the welfare of the USA and created a process for licensing private companies to develop, own, operate, and sell high-resolution data from Earth-observing satellites. Two years later four licences for one-meter systems were granted, and currently the first satellite, IKONOS, is in space. This innovation promises to set off an explosion in the amount and use of high resolution image data.

High-resolution imaging requires a change in instrument design to a pushbroom and large telescope, as well as a new spacecraft design. In contrast to the medium-resolution satellites, high-resolution systems have limited multispectral coverage, or even just panchromatic capabilities. They do have extreme pointing capabilities to increase their potential coverage. The pointing capability can also be used for last minute reprogramming of the satellite in case of cloud cover.

The private sector has shown an almost exclusive interest in high-resolution systems. Obviously, it is believed that these systems represent the space capability needed to create commercially valuable products. On the other hand, pure commercial remote sensing systems, with no government funding, implies a high risk, especially to data users. Most companies in the high-resolution business have a back-up satellite in store, in order to be able to launch a replacement satellite at short notice. But still, the loss of one satellite means a loss of millions of dollars, which may be considerable for a business just starting in this field. The characteristics of high-resolution satellites include a spatial resolution of less than 5 m, 1 to 4 spectral bands, a swath less than 100 km and a revisiting time of better than 3 days.

3.2.2 Overview sensors

An overview of high-resolution sensors to be discussed is given in Table 3.1.

Table 3.1: Typical high-resolution satellites.

Platform	Sensor	Spatial resolution	Multi-spectral	Swath width	Pointing capability	Revisit time
IRS-1C&D*	PAN	5.8 m	4 bands	70 km	±26°	5 days
Cosmos*	KVR-1000	~2 m	No	160 km	No	N/A
OrbView-3	PAN	1 m	4 bands	8 km	±45°	3 days
Ikonos 1	OSA	1 m	4 bands	11 km	±30°	1-3 days
QuickBird	QBP	1 m	4 bands	27 km	±30°	1-3 days
EROS A+	CCD	1.8 m	No	12.5 km		

3.2.3 IRS-1C and IRS-1D

Having been the seventh nation to successfully launch an orbiting remote sensing satellite in July 1980, India is pressing ahead with an impressive national programme aimed at developing launchers as well as nationally produced communications, meteorological and Earth resources satellites. The IRS-1C and 1D offer improved spatial and spectral resolution over the previous versions of the satellite, as well as on-board recording, stereo viewing capability and more frequent revisits. They carry three separate imaging sensors, the WiFS, the LISS, and the high-resolution panchromatic sensor.

The Wide Field Sensor (WiFS) provides regional imagery acquiring data with 800 km swaths at a coarse 188 m resolution in two spectral bands, visible (620-680 nm) and near infrared (770-860 nm), and is used for vegetation index mapping. The WiFS offers a rapid revisit time of 3 days.

The Linear Imaging Self-Scanning Sensor 3 (LISS-3) serves the needs of multispectral imagery clients, possibly the largest of all current data user groups. LISS-3 acquires four bands (520-590, 620-680, 770-860, and 1550-1750 nm) with

---

* IRS-1, Pan and Cosmos do not meet the strict definition of 'high resolution imagery', but is considered to be an example of this genre.

a 23.7 m spatial resolution, which makes it an ideal complement to data from the aging Landsat 5 Thematic Mapper (TM) sensor.

The most interesting of the three sensors is the panchromatic sensor with a resolution of 5.8 m. With its 5.8 m resolution, the IRS-1C and IRS-1D can cover applications that require spatial detail and scene sizes between the 10 m SPOT satellites and the 1 m systems. The PAN sensor is steerable up to plus or minus 26 degrees and thus offers stereo capabilities and a possible frequent revisit of about 5 days, depending on the latitude. Working together, the IRS-1C and 1D will also cater to users who need a rapid revisiting rate. IRS-1C was launched on 28 December 1995, IRS-1D on 28 September 1997. Both sensors have a 817 km orbit, are sun-synchronous with a 10:30 equator crossing, and a 24-day repeat cycle.

India will initiate a high-resolution mapping programme with the launch of the IRS-P5, which has been dubbed Cartosat-1. It will acquire 2.5 m resolution panchromatic imagery. There seem to be plans to futher improve the planned Cartosat-2 satellite to achieve 1 m resolution.

3.2.4 KVR-1000

Data from the Russian KVR-1000 camera, flown on a Russian Cosmos satellite, is marketed under the name of SPIN-2 (Space Information – 2 m). It provides high-resolution photography of the USA in accordance with a Russian-American contract. Currently SPIN-2 offers some of the world's highest resolution, commercially available satellite imagery. SPIN-2 panchromatic imagery has a resolution of about 2 m. The data is single band with a spectral range between 510 and 760 nm. Individual scenes cover a large area of 40 km by 180 km. Typically, the satellite is launched and takes images for 45 days, before it runs out of fresh film; the last mission was in February-March 1998. The KVR-1000 is in a low-earth orbit and provides 40 x 160 km scenes with a resolution.

3.2.5 OrbView-3

OrbView-3 will produce 1 m resolution panchromatic and 4 m resolution multispectral imagery. OrbView-3 is in a 470 km sun-synchronous orbit with a 10:30 equator crossing. The spatial resolution is 1 m for a swath of 8 km and a 3 day revisit time. The panchromatic channel covers the spectral range from 450 nm to 900 nm. The four multispectral channels cover 450–520 nm, 520–600 nm, 625–695 nm, and 760–900 nm respectively. The design lifetime of the satellite is 5 years. In Europe, Spot Image will have the exclusive right to sell the imagery of OrbImage's planned OrbView-3 and OrbView-4 satellites. OrbView-3 and OrbView 4 are planned to be launched in 2001.

3.2.6 Ikonos

The Ikonos satellite system was initiated as the Commercial Remote Sensing System (CRSS). The satellite will routinely collect 1 m panchromatic and 4 m multispectral imagery. Mapping North America's largest 100 cities is an early priority. The sensor OSA (Optical Sensor Assembly) features a telescope with a 10 m focal length (folded optics design) and pushbroom detector technology. Simultaneous imaging in the panchromatic and multispectral modes is provided. A body pointing technique of the entire spacecraft permits a pointing capability of ±30° in any direction. Ikonos is in a 680 km, 98.2°, sun-synchronous orbit with a 14 days repeat cycle and a 1–3 day revisit time. The sensor has a panchromatic spectral band with 1 m resolution (0.45–0.90) and 4 multispectral bands (0.45–0.52, 0.52–0.60, 0.63–0.69, 0.76–0.90) with 4 m resolution. The swath is 11 km.

3.2.7 QuickBird

QuickBird is the next-generation satellite of the EarlyBird satellite. Unfortunately, EarlyBird was lost shortly after launch in December 1997. Its follow-up QuickBird (QuickBird-1 was launched on 20 November 2000, and also failed). The system has a planed panchromatic channel (0.45–0.90) with 1 m resolution at nadir and four multispectral channels (0.45–0.52, 0.53–0.59, 0.63–0.69, 0.77–0.90) with 4 m resolution.

3.2.8 Eros

Eros (12.5 km swath) is the result of a joint venture between the US and Israel. The Eros A+ satellite will have a resolution of about 1.8 m. The follow-up satellite Eros B will have a resolution of about 80 cm.

EROS satellites are light, low earth orbiting, high resolution satellites. There are two classes of EROS satellite, A and B. EROS A1 and A2 will weigh 240 kg at launch and orbit at an altitude of 480 km. They will each carry a camera with a focal plane of CCD (Charge Coupled Device) detectors with more than 7,000 pixels per line. The expected lifetime of EROS A satellites is at least 4 years. EROS B1-B6 will weigh under 350 kg at launch and orbit at an altitude of 600 km. They carry a camera with a CCD/TDI (Charge Coupled Device/Time Delay Integration) focal plane that enables imaging even under weak lighting conditions. The camera system provides 20,000 pixels per line and produces an image resolution of 0.82 m. The expected lifetime of EROS B satellites is at least 6 years.

EROS satellites will be placed in a polar orbit. Both satellites are sun-synchronous. The light, innovative design of the EROS satellites allows for a great degree of platform agility. Satellites can turn up to 45 degrees in any direction as they orbit, providing the power to take shots of many different areas during the same pass. The satellites' ability to point and shoot their cameras also allows for stereo imaging during the same orbit. The satellites will be launched using refurbished Russian ICBM rocket technology, now called Start-1. Satellites will be launched from 2000-2005; EROS-A1 was launched on 5 December 2000.

3.2.9 Applications and perspectives

Satellite images have traditionally been used for military surveillance, to search for oil and mineral deposits, infrastructure mapping, urban planning, forestry, agriculture and conservation research. Agricultural applications may benefit from the increased resolution. The health of agricultural crops can be monitored by analyzing images of near-infrared radiation. Known as 'precision agriculture', farmers are able to compare images one or two days apart and apply water, fertilizer or pesticides to specific areas of a field, based on coordinates from the satellite image, and a Global Positioning System (GPS). In forestry, individual trees could be identified and mapped over large areas (see Chapter 6 by Woodcock *et al.*). Geographic information systems (GIS) databases may be constructed using 1 m images, reducing reliance on out-of-date paper maps. Highly accurate elevation maps (or Digital Elevation Models – DEMs), may be also be developed from the images and added to the databases. Because they cover large areas, high-resolution satellite images could replace aerial photographs for certain types of detailed mapping; for example, gas pipeline routing, urban planning and real estate. This includes the use of high resolution imagery for three-dimensional drapes that can be used to visualize and simulate land-management activities.

3.3   HIGH SPECTRAL RESOLUTION SATELLITES

3.3.1 Historical overview

Imaging spectrometry satellites use a near-continuous radiance or reflectance to capture all spectral information over the spectral range of the sensor. Imaging spectrometers typically acquire images in a large number of channels (over 40), which are narrow (typically 10 to 20 nm in width) and contiguous (i.e., adjacent and not overlapping – see Figure 3.2). The resulting reflectance spectra, at a pixel scale, can be directly compared with similar spectra measured in the field, or laboratory. This capability promises to make possible entirely new applications and to improve the accuracy of current multispectral analysis techniques. The demand for imaging spectrometers has a long history in the geophysical field; aircraft-based experiments have shown that measurements of the continuous spectrum allow greatly improved mineral identification (Van der Meer and Bakker 1997). The first civilian airborne spectrometer data were collected in 1981 using a one-dimensional profile spectrometer developed by the Geophysical Environmental Research Company. These data comprised 576 channels covering the 4 to 2.5 μm wavelength range (Chiu and Collins 1978). The first imaging device was the Fluorescence Line Imager (FLI; also known as the Programmable Line Imager, PMI) developed by Canada's Department of Fisheries and Oceans in 1981. The Airborne Imaging Spectrometer (AIS), developed at the NASA Jet Propulsion Laboratory was operational from 1983 onward. This instrument acquired data in 128 spectral bands in the range of 1.2–2.4 μm. with a field-of-view of 3.7 degrees resulting in images of 32 pixels width (Vane and Goetz 1988). A later version of the instrument, AIS-2, covered the 0.8–2.4 μm region acquiring images 64 pixels wide (LaBaw 1987). In 1987 NASA began operating the Airborne Visible/Infrared Imaging

Spectrometer (AVIRIS; Vane *et al.* 1993). AVIRIS was developed as a facility that would routinely supply well-calibrated data for many different purposes. The AVIRIS scanner simultaneously collects images in 224 contiguous bands resulting in a complete reflectance spectrum for each 20 by 20 m. pixel in the 0.4 to 2.5 μm region with a sampling interval of 10 nm (Goetz *et al.* 1983; Vane and Goetz 1993). The field-of-view of the AVIRIS scanner is 30 degrees resulting in a ground field-of-view of 10.5km. Private companies now recognize the potential of imaging spectrometry and have built several sensors for specific applications. Examples are the GER imaging spectrometer (operational in 1986), and the ITRES CASI that became operational in 1989. Currently operational airborne instruments include the NASA instruments (AVIRIS, TIMS and MASTER), the DAIS instrument operated by the German remote sensing agency DLR, as well as private companies such as HyVISTA who operate the HyMAP scanner or the Probe series of instruments operated by Earth Search Sciences, Inc.

Imaging Spectroscopy is the acquisition of images where for each spatial resolution element in the image a spectrum of the energy arriving at the sensor is measured. These spectra are used to derive information based on the signature of the interaction of matter and energy expressed in the spectrum. This spectroscopic approach has been used in the laboratory and in astronomy for more than 100 years, but is a relatively new application when images are formed from aircraft or spacecraft.

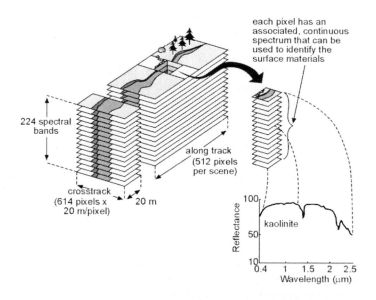

Figure 3.2: Concept of imaging spectroscopy.

3.3.2 Overview hyperspectral imaging sensors

An overview of imaging spectrometry sensors that are discussed here is given in Table 3.2.

Table 3.2: Some imaging spectrometry satellites.

Platform	Sensor	Spatial resolution	Spectral bands	Spectral range (μm)	Swath width	Revisit time
ENVISAT-1	MERIS	300 m	15			
EOS-AM1	ASTER	15-90 m	14	0.52–11.65	60 km	
Orbview 4		8 m	200	0.45–2.5	5 km	3 days
NMP/EO-1	Hyperion	30 m	220	0.4–2.5	7.5	
	LAC	250 m	256	0.9–1.6	185 km	
Aries-1		30 m	64	0.4–1.1	15 km	7 days
			64	2.0–2.5		

3.3.2.1 ENVISAT-1

The European Space Agency (ESA) is developing two spaceborne imaging spectrometers: The Medium Resolution Imaging Spectrometer (MERIS) and the High Resolution Imaging Spectrometer (HRIS); now renamed to PRISM, the Process Research by an Imaging Space Mission (Posselt *et al.* 1996). MERIS, currently planned as payload for the satellite Envisat-1 to be launched in 2002, is designed mainly for oceanographic application and covers the 0.39–1.04 μm wavelength region with 1.25 nm bands at a spatial resolution of 300 m or 1200 m. (Rast and Bézy 1995). PRISM, currently planned for Envisat-2 to be launched around the year 2003, will cover the 0.4–2.4 μm wavelength range with a 10 nm contiguous sampling interval at a 32 m ground resolution.

3.3.2.2 EOS-AM1

The EOS (Earth Observing System) is the centerpiece of NASA's Earth Science mission. The EOS AM-1 satellite, later renamed to Terra, is the main platform that was launched on 18 December 1999. It carries five remote sensing instruments (including MODIS and ASTER). EOS-AM1 orbits at 705 km, is sun-synchronous with a 10:30 equator crossing and a repeat cycle of 16 days. ASTER (the Advanced Spaceborne Thermal Emission and Reflectance Radiometer) has three bands in the visible and near-infrared spectral range with a 15 m spatial resolution, six bands in the short wave infrared with a 30 m spatial resolution, and five bands in the thermal infrared with a 90 m spatial resolution. The VNIR and SWIR bands have a spectral resolution in the order of 10 nm. Simultaneously, a single band in the near-infrared will be provided along track for stereo capability. The swath width of an image will be 60 km with 136 km crosstrack and a temporal resolution of less than 16 days. Also on the EOS-AM1, the Moderate resolution imaging spectroradiometer (MODIS) is planned as a land remote sensing instrument with high revisting time. MODIS is mainly designed for global change research (Justice *et al.*, 1998).

ASTER carries three telescopes: VNIR 0.56, 0.66, 0.81 µm; SWIR 1.65, 2.17, 2.21, 2.26, 2.33, 2.40 µm; TIR 8.3, 8.65, 9.10, 10.6, 11.30 µm with spatial resolutions of VNIR 15 m, SWIR 30 m, TIR 90 m.

3.3.2.3   OrbView 4

OrbView-4 will be the successor of the OrbView-3 high-resolution satellite. As with OrbView-3, OrbView-4's high-resolution camera will acquire 1 m resolution panchromatic and 4 m resolution multispectral imagery. In addition, OrbView-4 will acquire hyperspectral imagery. The sensor will cover the 450 to 2500 nm spectral range with 8 m nominal resolution and a 10 nm spectral resolution in 200 spectral bands. The data available to the public will be resampled to 24 m. The 8 m data will only be used for military purposes. OrbView-4 will be launched on 31 March 2001. The satellite will revisit each location on Earth in less than three days with an ability to turn from side-to-side up to 45 degrees from a polar orbital path.

3.3.2.4   EO-1

NASA's New Millennium Program Earth Observer 1 (NMP/EO-1; see Table 3.3) is an experimental satellite carrying three advanced instruments as a technology demonstration (EO-1 is now called Earth Observing-1). It carries the Advanced Land Imager (ALI), which will be used in conjunction with the ETM+ sensor (see Landsat 7 below for a comparison of the two sensors). Next to the multispectral instrument it carries two hyperspectral instruments, the Hyperion and the LEISA Atmospheric Corrector (LAC). The focus of the Hyperion instrument is to provide high-quality calibrated data that can support the evaluation of hyperspectral technology for spaceborne Earth observing missions. It provides hyperspectral imagery in the 0.4 to 2.5 µm region at continuous 10 nm intervals. Spatial resolution will be 30 m. The LAC is intended to correct mainly for water vapour variations in the atmosphere using the information in the 890 to 1600 nm region at 2 to 6 nm intervals. In addition to atmospheric monitoring, LAC will also image the Earth at a spatial resolution of 250 m. The imaging data will be cross-referenced to the Hyperion data where the footprints overlap. The EO-1 was successfully launched on 21 November 2000.

Table 3.3: Characteristics of EO-1.

	Hyperion	LAC
Spectral range	0.4–2.5 m	0.9–1.6 m
Spatial resolution	30 m	250 m
Swath width	7.5 km	185 km
Spectral resolution	10 nm	2–6 nm
Spectral coverage	continuous	continuous
Number of bands	220	256

3.3.2.5 Aries-1

Aries-1 is a purely Australian initiative to build a hyperspectral satellite, mainly targeted at geological applications for the (Australian) mining business. The ARIES-1 will be operated from a 500 km sun-synchronous orbit. The system will have a VNIR and SWIR hyperspectral, and PAN band setting with 128 bands in the 0.4 – 1.1 µm and 2.0 – 2.5 µm regions. The PAN band will have 10 m resolution, the hyperspectral bands will have 30 m resolution. The swath width is 15 km with a revisit time of 7 days.

3.3.3 Applications and perspectives

The objective of imaging spectrometry is to measure quantitatively the components of the Earth from calibrated spectra acquired as images for scientific research and applications. In other words, imaging spectrometry will measure physical quantities at the Earth's surface such as upwelling radiance, emissivity, temperature and reflectance. Based upon the molecular absorptions and constituent scattering characteristics expressed in the spectrum, the following objectives will be researched and solution found to:

- Detect and identify the surface and atmospheric constituents present
- Assess and measure the expressed constituent concentrations
- Assign proportions to constituents in mixed spatial elements
- Delineate spatial distribution of the constituents
- Monitor changes in constituents through periodic data acquisitions
- Simulate, calibrate and intercompare sensors.

Through measurement of the solar reflected spectrum, a wide range of scientific research and application is being pursed using signatures of energy, molecules and scatterers in the spectra measured by imaging spectrometers. Atmospheric science includes the use of hyperspectral sensors for the prediction of various constituents such as gases and water vapour. In ecology, some use has been made of the data for quantifying photosynthetic and non-photsynthetic constituents. In geology and soil science, the emphasis has been on mineral mapping to guide in mineral prospecting. Water quality studies have been the focus of coastal zone studies. Snow cover fraction and snow grain size can be derived from hyperspectral data. Review papers on geological applications can be found in van der Meer (1999). Cloutis (1996) provides a review of analytical techniques in imaging spectrometry while Van der Meer (2000) provides a general review of imaging spectrometry. Clevers (1999) provides a review of applications of imaging spectrometry in agriculture and vegetation sciences.

3.4   HIGH TEMPORAL RESOLUTION SATELLITES

3.4.1 Low spatial resolution satellite systems with high revisiting time

Typically, these satellites (Table 3.4) have a spatial resolution larger than 100 m. They trade reduced spatial and spectral resolution against high frequency visits. A global system of geo-stationary and polar orbiting satellites is used to observe global weather. Other satellites are used for oceanography, and for mapping phenomena on a continental or even global scale. Typical low-resolution satellite systems have a spatial resolution of 100 m or lower, few (3–7) spectral bands, large (>500 km) swath width and daily revisit capability.

Table 3.4: A selection of low-resolution satellites with high revisiting time.

Platform	Orbit	Sensor	Spatial resolution	Spectral bands	Swath width	Revisit time
Meteosat	GEO	VISSR	2.5 km	3	Earth Disc	30 min
NOAA	Polar	AVHRR	1	7	3000 km	Daily
Resurs-O1	Sun-sync	MSU-SK1	200–300 m	4	760 km	3–5 days
SeaStar	Sun-sync	SeaWiFS	1.1 km	8	2800 km	Daily

3.4.1.1   Meteosat

Meteosat 1 was the first European meteorological geo-stationary satellite. Meteosat 5 is currently the primary satellite, with Meteosat 6 as standby. Meteosat is controlled by Eumetsat, an international organization representing 17 European states. Meteosat Second Generation (MSG) will appear in the year 2000, together with the first polar orbiting Metop satellite. Meteosat is in a geo-stationary orbit at 0° longitude. The sensor has spectral bands at 0.5–0.9 μm (VIS), 5.7–7.1 μm (WV), and 10.5–12.5 μm (TIR) with spatial resolutions of 2.5 km VIS and WV and 5 km TIR. The revisit time is 30 minutes.

3.4.1.2   NOAA

The NOAA satellite program, designed primarily for meteorological applications, has evolved over several generations of satellites (TIROS, ESSA, TIROS-M, and TIROS-N, to NOAA-KLM series), starting with TIROS-1 through to the most recent NOAA-15. These satellites have provided different instruments for measuring the atmosphere's temperature and humidity profiles, the Earth's radiation budget, space environment, instruments for distress signal detection (search and rescue), instruments for relaying data from ground-based and airborne stations, and more.

For Earth observation the most interesting instrument is the Advanced Very High Resolution Radiometer (AVHRR) scanner. The AVHRR scans the Earth in five spectral bands: band 1 in the visible red around 0.6 μm, band 2 in the near infrared around 0.9 μm, band 3 in the mid-wave infrared around 3.7 μm, and band 4 and 5 in the thermal infrared around 11 and 12 μm respectively. This combination of bands makes the AVHRR suitable for a wide range of applications, from measurement of cloud cover, to sea surface temperature, vegetation, land and sea ice. The disadvantage of the AVHRR is its coarse resolution of about 1 km at nadir. But the major benefit of the AVHRR lies in its high temporal frequency of coverage.

The NOAA satellites are operated in a two-satellite system. Both satellites are in a sun-synchronous orbit, one satellite will always pass around noon and midnight, the other always passing in the morning and in the evening. The AVHRR sensors have an extreme field of view of 110°, and together they give a global coverage each day! Every spot on Earth is imaged at least twice each day, depending on latitude. It is *the* instrument for observation of phenomena on a global scale. Owing to its frequent revisit time, it is being used for many monitoring projects on a regional scale.

The imagery of the AVHRR is also known by other names. The HRPT (High Resolution Picture Transmission) is the digital real-time reception of the imagery by a ground station. There are over 500 HRPT receiving stations registered by the World Meteorological Organization (WMO) worldwide. The satellite can also be programmed to record a number of images. Such images, although having the same characteristics as HRPT, are called LAC (Local Area Coverage). Next to the 1 km resolution LAC, the satellite can resample the data on the fly to 4 km resolution GAC (Global Area Coverage). Finally, two bands of 4 km resolution imagery are transmitted by an analogue weather fax signal from the satellite, which can be received by relatively simple and low-cost equipment. This is called the APT (Automatic Picture Transmission). Two excellent sources of information on NOAA are Cracknell (1997) and D'Souza *et al.* (1996).

NOAA-14 (since 30 Dec 1994) and NOAA-15 (since 13 May 1998) are in a 850 km, 98.9°, sun-synchronous (afternoon or morning) orbit. The spatial resolution is 1 km at nadir, 6 km at limb of sensor. Spectral bands include band 1 at 580-680 nm, band 2 at 725-1100 nm, band 3 at 3.55-3.93 μm, band 4 at 10.3-11.3 μm and band 5 at 11.4-12.4 μm. The revisit time is 2-14 times per day, depending on latitude. NOAA-16 was launched on 21 September 2000.

3.4.1.3 Resurs-O1

Launched on 10 July 1998, the Resurs-O1#4 is the fourth operational remote sensing satellite in the Russian Resurs-O1 series. Maybe it is not altogether fair to list the Russian Resurs under the low-resolution category as it is actually equivalent to the US Landsat. But the satellite is best known for its relatively cheap large coverage images of the MSU-SK conical scanner. There are only two receiving stations located in Russia and in Sweden. The Swedish Space Corporation Satellitbild also processes and distributes the images. With a swath of 760 km and resolution of about 250 m Resurs fills the gap between the 1 km resolution NOAA images and the 30 m resolution Landsat images. Resurs-O1 is in a 835 km, 98.75°, sun-synchronous orbit. The sensor has spectral bands at 0.5–0.6 μm, 0.6–0.7 μm, 0.7–0.8 μm, 0.8–1.1 μm and 10.4–12.6 μm with a 30 m (MSU-E) and 200–300 m (MSU-SK) spatial resolution. The swath width is 760 km with a 3–5 day revisit time.

3.4.1.4   OrbView-2, a.k.a. SeaStar/SeaWiFS

Launched on 1 August 1997, SeaStar delivers multispectral ocean-colour data to NASA until 2002. This is the first time that the US Government has purchased global environmental data from a privately designed and operated remote sensing satellite. SeaStar carries the SeaWiFS Sea-viewing Wide Field Sensor, which is a next generation of the Nimbus 7's Coastal Zone Color Scanner (CZCS). SeaWiFS measures ocean surface-level productivity of phytoplankton and chlorophyll. However, SeaStar was originally designed for ocean colour but later changed to be able also to measure the higher radiances from land. Thus, it provides a more environmentally stable vegetation index than the one derived from NOAA's AVHRR, which is inaccurate under hazy atmospheric conditions because of its single visible and near infrared channels. Band 1 looks at gelbstoffe, bands 2 and 4 at chlorophyll, band 3 at pigment, band 5 at suspended sediments. Bands 6, 7 and 8 look at atmospheric aerosols, and are provided for atmospheric corrections.

   Orbview-2 is in a 705 km, 98.2°, sun-synchronous, equator crossing (at noon) orbit. The spatial resolution of the data is 1.1 km, the swath width is 2800 km with a 1 day revisit time. Spectral bands of the system include: 402–422 nm, 433–453 nm, 480–500 nm, 500–520 nm, 545–565 nm, 660–680 nm, 745–785 nm, 845–885 nm.

3.4.2 Medium spatial resolution satellite systems with high revisiting time

These satellites (Table 3.5) all have medium area coverage, a medium spatial resolution, a moderate revisit capability, and multispectral bands characteristic of the current Landsat and Spot satellites. The scale of the images of these satellites makes them especially suited for land management and land-use planning for extended areas (regions, countries, continents). Most of these medium-resolution satellites are in a sun-synchronous orbit.

   Characteristics of medium-resolution and satellites with high revisiting time include:

* Spatial resolution between 10 m and 100 m
* 3 to 7 spectral bands
* Swath between 50 km and 200 km
* Incidence angles
* Revisit 3 days and more.

Table 3.5: Some medium-resolution satellites.

Platform	Sensor	Spatial resolution	Spectral bands	Swath width	Pointing capability	Revisit time
Landsat 4&5	TM	30 m	7	185 km	No	16 days
Landsat 7	ETM+	15 m (PAN)	8	185 km	No	16 days
Spot 1-3	HRV	10 m (PAN)	3	60 km	±27°	4–6 days
Spot 4	HRVIR	10 m (PAN)	4	60 km	±27°	4–6 days
Resource21	M10	10 m (PAN)	6	205 km	±30°	3–4 days

3.4.2.1 Landsat

Earth observation is often associated with the Landsat satellites, having been in operation since 1972, while Landsat 5 has been in operation for 15 years! The Landsat satellites were developed in the 1960s. Landsats 1, 2 and 3 were enhanced versions of the Nimbus weather and research satellites, and originally known as the Earth Resources Technology Satellites (ERTS). The moment Landsat 1 became operational its images were regarded as sensational by early investigators. The quality of the images led to the information being put to immediate practical use. It became clear that they were directly relevant to the management of the world's food, energy and environment. The Landsat satellites have flown the following sensors:

- The Return Beam Vidicon (RBV) Camera
- Multi-spectral Scanner (MSS)
- Thematic Mapper (TM).

Landsat 4 was launched on 16 July 1982, and a failing power system on Landsat 4 prompted the launch of Landsat 5 two years later on 1 March 1984. The loss of Landsat 6 in October 1993 was a severe blow to the system. Both Landsat 4 and 5 suffer from degrading sub-systems and sensors and are expected to fail any moment. At present, Landsat 7 is in operation.

The characteristics of Landsats 4 and 5 include:

- Operational: Landsat 5 (since 1 March 1984!)
- Orbit: 705 km, 98.2°, sun-synchronous 09:45 AM local time equator crossing
- Repeat cycle: 16 days
- Sensor: Thematic Mapper (TM), electro-mechanical oscillating mirror scanner
- Spatial resolution TM: 30 m (band 6: 120 m)
- Spectral bands TM (μm): band 1 0.45–0.52; band 2 0.52–0.60; band 3 0.63–0.69; band 4 0.76–0.90; band 5 1.55–1.75; band 6 10.4–12.50; band 7 2.08–2.35
- Field of view (FOV): 15°, giving 185 km swath width.

Characteristics of Landsat 7 include:

- Operational: Landsat 7 Orbit: 705 km, 98.2°, sun-synchronous 10 AM local time equator crossing
- Repeat cycle: 16 days
- Sensor: Enhanced Thematic Mapper+ (ETM+), electro-mechanical oscillating mirror scanner; imagery can be collected in low- or high-gain modes; high gain doubles the sensitivity
- Spatial resolution: 30 m (PAN: 15 m, band 6: 60 m)
- Spectral bands (μm): band 1 0.45–0.52; band 2 0.52–0.60; band 3 0.63–0.69; band 4 0.76–0.90; band 5 1.55–1.75; band 6 10.4–12.50; band 7 2.08–2.35; band 8 (PAN) 0.50–0.90
- Field of view (FOV): 15°, 185 km
- Downlink: X-band, 2x150 Mbit/s, 300 Mbit/s playback
- Onboard recorder: 375 Gbit Solid State Recorder for about 100 ETM+ scenes.

3.4.2.2   SPOT 1/2/3 (Système Pour l'Observation de la Terre)

First named Système Probatoire d'Observation de la Terre (Test System for Earth Observation), but later renamed to Système Pour l'Observation de la Terre (System for Earth Observation), the Spot system has been in operation since 1986. The Spot satellites each carry two identical HRV (High-Resolution Visible) sensors, consisting of CCD (Charge-Coupled Device) linear arrays. Essentially, all the points of one line are imaged at the same time by the many detectors of the linear array. The Spot sensors can be tilted from the normal downward viewing mode by plus or minus 27 degrees, which offers the possibility to view objects from two different sides. These stereo images can be used to determine the height of objects, or even the height of the terrain. Spot 1 was already put in standby mode, but reactivated after Spot 3 failed on 14 November 1996. Spot 3 ran out of power when an incorrect series of commands was sent to the satellite, and could not be recovered. An important addition to Spot 4 is the VEGETATION instrument. With resolution of 1.15 km at nadir, and a swath width of 2,250 kilometers, the VEGETATION instrument will cover almost all of the globe's landmasses while orbiting the Earth 14 times a day. Characteristics of Spot 1/2/3 include:

- Orbit: 832 km, 98.7°, sun-synchronous 10:30 AM local time equator crossing
- Repeat cycle: 26 days
- Sensor: 2xHRV (High Resolution Visible), pushbroom linear CCD array
- Spatial resolution: MS mode 20 m, PAN 10 m
- Spectral bands MS (nm): band 1 500–590; band 2 610–680; band 3 790–890
- Panchromatic mode: 510–730 nm
- Swath width: 60 km
- Steerable: ±27° left and right from nadir
- Of-nadir Revisit time: 4–6 days.

Characteristics of Spot 4 include:

- Orbit: same as SPOT 1-2-3
- Sensors: 2xHRVIR (High-Resolution Visible and Infrared), pushbroom linear CCD array, VEGETATION
- Spatial resolution: MS mode 20 m, PAN 10 m, VEGETATION 1.15 km
- Spectral bands MS (nm): band 1 500–590; band 2 610–680; band 3 790–890; band 4 1.58–1.75 μm
- Panchromatic mode: 610–680 nm (same as MS band 2!)
- VEGETATION bands: band 1 0.43, bands 2/3/4 same as HRVIR
- VEGETATION swath: 2250 km.

3.4.2.3 Resource21

Resource21 is the name of a commercial remote sensing information services and services company based in the US. Resource21 will combine satellite and aircraft remote sensing to provide twice-weekly information products within hours of data collection, based on 10 m resolution multispectral. The complete system will consist of four satellites. All areas on Earth will be visited by one of the satellites every three or four days, resulting in a revisit twice a week. In other words, the constellation will give a global coverage every three or four days at 10 m resolution. The main application areas of Resource21 are agriculture ('precision farming'), and environment and natural resource monitoring.

Resource21 is at a 740 km, 98.6°, sun-synchronous, 10:30 crossing at ascending node orbit. The sensor has spectral bands (μm) at 0.45–0.52 blue, 0.53–0.59 green, 0.63–0.69 red, 0.76–0.90 NIR, 1.55–1.68 SWIR, and 1.23–1.53 cirrus clouds with a 10 m resolution in the VNIR, a 20 m resolution in the SWIR, and a 100 m+ resolution for the cirrus band. The swath width is 205 km with a 3–4 day revisit time. The status of the Resource21 programme is unclear. It has been on hold for some time because of budget considerations, but Boeing continues to work towards a realization of the programme.

3.5 RADAR

3.5.1 Historical overview

RADAR (Radio Detection And Ranging) remote sensing has been used operationally since the 1960s. The technique uses the microwave and radio part of the spectrum, with frequencies between 0.3 GHz and 300 GHz roughly corresponding to wavelengths between 1 mm and 1 m, with wavelengths between 0.5 and 50 cm being widely utilized. Table 3.6 summarizes the available wavelengths, their usage and the availability in terms of sensors.

Table 3.6: Radar bands used in Earth observation, with corresponding frequencies.
Sources: Hoekman (1990) and van der Sanden (1997).

Band	Frequency (GHz)	Available in
Ka	35.5 – 35.6	Airborne sensors
K	24.05 – 24.25	Airborne sensors
Ku	17.2 – 17.3	Airborne sensors
Ku	13.4 – 14.0	Airborne sensors
X	9.50 – 9.80	Airborne sensors
X	8.55 – 8.65	Airborne sensors
C	5.25 – 5.35	Airborne sensors, ERS-1, -2, RADARSAT
S	3.1 – 3.3	ALMAZ
L	1.215 – 1.3	Airborne sensors, SEASAT, JERS-1 (out of use)
P	0.44 (central frequency)	Airborne sensors

The first airborne systems were called SLAR (Side Looking Airborne Radar), as the instrument looked at an area that was not directly under, but on one side of, an aircraft. These radar systems are also called RAR (Real Aperture Radar), as the real length of the antenna was used (Hussin *et al.* 1999).

Further technical developments made it possible to increase the antenna length virtually by using the Doppler effect. These systems are called SAR (Synthetic Aperture Radar). This new technique made it possible to increase the spatial resolution without increasing the antenna length (with the sensor flying at the same height) or to mount the sensor on a satellite without loosing too much of the spatial resolution. SAR systems also look at one side of the aircraft or the satellite, however, the term 'SLAR' is never used to describe these radar systems.

The latest development in radar technology is the so-called active phased array antenna, presently only operational in the airborne version in the PHARUS system (PHased ARray Universal SAR), but in future it may be also available in spaceborne systems (i.e., the ASAR instrument of the planned ENVISAT satellite). The advantage of this system is that it is relatively small and lightweight.

Most imaging radar systems use an antenna that generates a horizontal waveform, referred to as horizontal polarization. After backscattering from an object, a portion of the returned energy may still retain in the same polarization as that of the transmitted signal. However, some objects tend to depolarize (the vibration of the wave deviates from its original direction) a portion of the microwave energy. The portion of the signal that is depolarized is a function of surface roughness and structural orientation. The switching mechanism in the emitting and receiving system, alternating between two receiver or emitter channels, permits both the horizontal and the vertical signals to be emitted or echoes to be received and recorded. The notation HH is used to denote that the signal is horizontally emitted and the echo horizontally received, while HV indicates horizontal emission and vertical reception. In the same way the notations VV and VH are used. HH and VV represent like polarization, while HV and VH are called cross-polarization return. HV and VH tend to produce similar results, therefore they are not used simultaneously. The like, or cross-polarization is an important factor when considering the orientation of the ground objects or their geometric properties, e.g. a vertical stump of a tree returns more vertical than horizontally polarized signals.

Radar remote sensing is now used with interferometry. For example, the use of SAR as an interferometer, the so-called SAR interferometry (InSAR; Massonnet *et al.*, 1994), is technically difficult due to stringent requirements for stability of the satellite orbit. Interferometry coupled with satellite SAR offers the possibility to map the Earth's land and ice topography and to measure small displacements over large temporal and spatial scales with subcentimeter accuracy, independent of sun illumination and cloud coverage. Interferometric SAR makes use mainly of the phase measurements in two or more SAR images of the same scene, acquired at two different moments and at two slightly different locations. By interference of the two images, very small slant range of the same surface can be inferred. These slant range changes can be related to topography and/or surface deformations.

The most important drawback of radar images is that the reflections of radar signals are very complex functions of the physical and structural properties, as well as water content of the target objects. This means that the interpretation of radar images is completely different than the interpretation of optical images.

Characteristics of radar satellites include:

- Spatial resolution between 10 m to 100 m
- Single or multiple (up to 3) radar frequencies
- Single or multiple polarization modes (HH, VV, HV, VH)
- Wavelengths/frequencies (see Table 3.6).

3.5.2 Overview of sensors

3.5.2.1 ERS-1 and 2

The European Remote Sensing (ERS 1) satellite has three primary all-weather instruments providing systematic, repetitive global coverage of ocean, coastal zones and polar ice caps, monitoring wave height and wavelengths, wind speed and direction, precise altitude, ice parameters, sea surface temperature, cloud top temperature, cloud cover, and atmospheric water vapour content. ERS-1 was launched on 17 July 1991, it went out of service on 10 March 2000 due to a failure of the attitude control system. ERS-2 was launched on 21 April 1995.

The Active Microwave Instrument (AMI) can operate as a wind scatterometer or SAR. The Along-Track Scanning Radiometer & Microwave Sounder (ATSR-M) provides the most accurate sea surface temperature to date. The Radar Altimeter (RA) measures large-scale ocean and ice topography and wave heights. In addition to these instruments, the ERS 2 also carried a Global Ozone Monitoring Experiment (GOME) to determine ozone and trace gases in troposphere and stratosphere.

The ERS 2 satellite has provided continuity of data until the launch of Envisat 1. ERS 1 and 2 operated simultaneously from 15 August to May 1996, the first time that two identical civil SARs worked in tandem. The orbits were carefully phased to provide 1 day revisits, allowing the collection of interferometric SAR image pairs and improving temporal sampling. Although still working perfectly, a lack of funding required ERS 1 to be put on standby from May 1996.

ERS is a C-band radar system with a 100 km swath and a VV polarization. The orbital parameters include a 771x797 km, 98.55°, sun-synchronous orbit with a 168-day repeat cycle.

3.5.2.2 *Radarsat*

Radarsat was launched on 4 November 1995. The Boeing Company will launch Canada's Radarsat-2 Earth-observation satellite, synthetic aperture radar (SAR) system in 2003. Radarsat-2, follow-on to Radarsat-1, is a jointly funded programme between the Canadian Space Agency (CSA) and MacDonald Dettwiler. It is part of a public-private sector partnership and a further step towards commercialization of the spaceborne radar imaging business.

Designed, constructed, launched, and operated by the CSA, Radarsat is a commercial radar remote sensing satellite dedicated to operational applications. Its C-band SAR has seven different modes of 10 to 100 m resolution and 50 to 500 km swath widths combined with 25 beam positions ranging from 10 to 60 degrees incidence angles. Thus, a wide variety of products can be offered. The Radarsat system is designed to operate with no backlog, so that all imagery can be processed and distributed within 24 hours and be available to a customer within 3 days of acquisition.

3.5.2.3 *Envisat*

In 2002 the European Space Agency will launch Envisat-1, an advanced polar-orbiting Earth observation satellite, which will provide measurements of the atmosphere, ocean, land, and ice. Envisat-1, will be placed in a 800 km, 98.55°, orbit with a crossing at the equator at 10:00 and a 35 day repeat cycle, are a large satellite carrying a substantial number of sensors. The most important sensors for land applications are the Advanced Synthetic Aperture Radar (ASAR), the Medium Resolution Imaging Spectrometer (MERIS), and the Advanced Along-Track Scanning Radiometer (AATSR).

An ASAR, operating at C-band, ensures continuity with the image mode (SAR) and the wave mode of the ERS-1/2 AMI. It features enhanced capability in terms of coverage, range of incidence angles, polarization, and modes of operation. The ERS-1 and 2 could only be switched on for 10 minutes each orbit, while the ASAR can take up to 30 minutes of high resolution imagery each revolution.

In normal image mode the ASAR will generate high spatial resolution products (30 m) similar to the ERS SAR. It will image one of the seven swaths located over a range of incidence angles spanning from 15 to 45 degrees in HH or VV polarization. As well as cross-polarization modes (HV and VH). In addition to the 30 m resolution it offers wide swath modes, providing images of a wider strip (405 km) with medium resolution (150 m). The SAR system on Envisat, ASAR, is a C-band system with HH and VV, or HH/HV, or VV/VH modes of polarization. The spatial resolutions are 30 m or 150 m with a swath width of 56–100 km.

3.5.2.4 SIR-C/X-SAR

Together the Shuttle Imaging Radar C (SIR-C, built by JPL/NASA), and the X-band Synthetic Aperture Radar (X-SAR, jointly built by Germany and Italy) allow radar images to be collected at 3 different wavelengths (X, C, and L-band) and 4 different polarization modes (HH, VV, HV, and VH). The first mission was flown on board the Space Shuttle flight STS-59 in April 1994. The second mission was aboard STS-68 in September-October 1994. Roughly 20 per cent of the Earth was imaged at up to 10 m resolution. SIR-C and X-SAR have a spatial resolution of 25 m and use the X, C, and L-band frequencies with HH, VV, HV, VH, polarizations on SIR-C and VV for X-SAR. The swath width is 15–60 km (15–45 km X-SAR).

NASA has launched the Shuttle Radar Topography Mission (SRTM; launched on 11 February 2000) to map a digital elevation map with 16 m height accuracy at 30 m horizontal intervals. SRTM uses the SIR-C antenna working interferometrically with additional antenna on a 60 m long deployable mast. Approximately 80 per cent of the Earth's landmass (everything between 60° north and 56° south latitude) will be imaged in 1000 scenes.

3.5.2.5 JERS-1

JERS-1 is the Japanese Earth Resources Satellite, launched on 11 February 1992. In addition to the SAR system, it carries the OPS (Optical Sensor), which has 7 downward looking bands and one forward viewing band for stereo viewing and uses the L-band frequency with HH polarization. The spatial resolution is 18 m at a 75 km swadth width. The original design lifetime was 2 years, but JERS-1 operated until 1998, when a short-circuit in its solar panels immobilized it. The next Japanese radar system will be on the ALOS (Advanced Land Observing Satellite). ALOS will be launched in the summer of 2003.

An overview of spaceborne radar remote sensing sensors is given in Table 3.7.

Table 3.7: A selection of spaceborne radar remote sensing instruments.

Platform	Sensor	Spatial resolution	Frequency band	Polarization	Swath width	Incidence angle
ERS-1&2	AMI/S AR	30 m	C-band	VV	100 km	23° (up to 35°)
Radarsat-1	SAR	8–100 m	C-band	HH	50–500 km	10–60°
Envisat-1	ASAR	30 m	C-band	HH+VV, HH/HV,	56–100 km	17–45°
		150 m		VV/VH	405 km	
Space Shuttle	SIR-C	25 m	C + L band	HH,VV,HV, VH	15–60 km	20–55°
(2x in 1994)	X-SAR		X-band	VV	15–45 km	
JERS-1	SAR	18 m	L-band	HH	75 km	35°

3.5.3 Applications and perspectives

Although some passive radar systems exist, using only the radiation emitted or reflected by the Earth, most radar sensors are active sensors that emit a signal and receive the portion of that signal that is scattered back to the sensor. Therefore, active radar sensors are independent of solar radiation and can operate both day and night. Because they utilize longer wavelengths, radar remote sensing is not obstructed by clouds or by rain. Heavy rainstorms may affect the radar image, especially if a small wavelength was used.

Radar provides information that is different from that obtained from optical and near infrared (NIR) remote sensing. Optical/NIR remote sensing for vegetation is based on differential scattering and absorption by the chlorophyll and leaf area, structural characteristics and chemical composition. In contrast, the radar return signal, or backscatter, is determined by the structure and roughness of the canopy, the spatial distribution of the parts of the plants as well as the moisture content of the plants, and the soil surface characteristics, such as roughness and moisture content. The longer radar wavelengths can penetrate tree canopies and topsoil. The use of SAR as an interferometer opens the possibility of accurate measuring of terrain height and height differences in time leading to estimates of surface deformation with applications in tectonics, volcanology, ice sheet mapping and so on.

3.6 OTHER SYSTEMS

3.6.1 Altimetry

Altimeters use the ranging capability of radar to measure the surface topographic profile, by simply emitting a pulse and measuring the time elapsed between emission and reception. The height h carries a linear relationship with the time lapse t: h=ct/2 (Elachi, 1987). Altimeters have been flown on a number of spacecraft, including Seasat, ERS 1&2, and TOPEX/Poseidon. Seasat produced the first images of the topography of the ocean floor, as a dip in the ocean floor causes a dip in the water surface due to reduced gravitational pull. The ocean surface can vary by up to 150 m. Knowing the mean water level of the oceans, minor variations around the mean level can be measured. TOPEX/Poseidon measures ocean temperature by these changes. There appears to be a direct relationship between temperature and water height. The sea surface temperature reflects the temperature in the top few centimeters of water. The sea surface measured by altimetry is related to the temperature at all depths, as well as other parameters, such as the water salinity and ocean currents. Note that, in general, the movement of currents has a bigger effect on sea level (± 1 m) than heating and/or winds (± 12 cm). All spaceborne radar altimeters have been wide-beam systems, limited in accuracy by their pulse duration. Such altimeters are useful for smooth surfaces (oceans), but are ineffective over continental terrain with relatively high relief. A fundamental problem in narrow-beam spaceborne radar altimetry is the physical constraint of antenna size. Again, large antennas are required for small radar footprints.

3.6.2 Scatterometers/spectrometers

Scatterometers are radar sensors that provide the backscattering cross section of the surface area illuminated by the sensor. They are particularly useful in measuring the ocean backscatter in order to derive the near-surface wind vector. The strength of the radar backscatter is proportional to the amplitude of the water surface capillary and small gravity waves (Bragg scattering), which in turn is related to the wind speed and direction near the surface (Elachi 1987).

3.6.3 Lidar

Lidar (LIght Detection And Ranging) refers to laser-based remote sensing. Lidar uses principles very much like the ones used in radar. The main difference is in wavelength; radar uses microwaves; lidar uses visible light. In fact, the term Lidar is a generic term used for a variety of sensors operated by different concepts. Going into detail in each of them would be beyond the scope this chapter.

The main applications of Lidar are measuring distance (height), movement, and chemical composition. By measuring the intensity, polarization, and spectral properties of the return signal as a function of time, one can obtain information on the properties of the atmosphere.

3.7 INTERNET SOURCES

The following provides an overview of internet sources of sensors listed in this chapter and includes sources of information for further reading.

3.7.1 High spatial resolution satellite systems

IRS: http://www.nrsa.org/
KVR: http://www.spin-2.com/
Orbital image (Orbview satellites): http://www.orbimage.com/
Quickbird/Early bird: http://www.digitalglobe.com/company/satellites.html
Ikonos: http://www.spaceimage.com/
EROS: http://www.westindianspace.com/
EROS: http://www.imagesatintl.com/

3.7.2 High spectral resolution satellite systems

ENVISAT: http://envisat.estec.esa.nl/instruments/
EOS-AM1/TERRA: http://terra.nasa.gov/
EOS-PM1/Aqua: http://aqua.gsfc.nasa.gov/
Orbview-4 Orbital image (Orbview satellites): http://www.orbimage.com/
NMP/EO-1: http://eo1.gsfc.nasa.gov/
ARIES: http://www.aries-sat.com.au/

3.7.3 High temporal resolution satellite systems

Meteosat: http://www.esoc.esa.de/external/mso/meteosat.html
NOAA (institute): http://www.noaa.gov/
NOAA (RS data): http://www.nesdis.noaa.gov/
NOAA-POES: http://poes.gsfc.nasa.gov/
Resurs: http://resurs.satellus.se/
Orbview-2 Orbital image (Orbview satellites): http://www.orbimage.com/
SeaWifs: http://seawifs.gsfc.nasa.gov/SEAWIFS.html
Landsat-7 (at USGS): http://landsat7.usgs.gov/
Landat-7 (at NASA): http://landsat.gsfc.nasa.gov/index.htm
Landsat program: http://geo.arc.nasa.gov/sge/landsat/landsat.html
SPOT: http://www.spotimage.fr/

3.7.4 RADAR satellite systems

ERS satellites: http://earth.esa.int/l2/2/ersnewhome
Radarsat: http://radarsat.space.gc.ca/
ENVISAT ASAR: http://envisat.estec.esa.nl/instruments/asar/index.html

SIR: http://southport.jpl.nasa.gov/
JERS (at NASDA): http://www.nasda.go.jp/
ALOS: http://www.eorc.nasda.go.jp/ALOS/
SRTM: http://www.jpl.nasa.gov/srtm/

3.7.5 General sources of information

3.7.5.1 General remote sensing sites

Good points for remote sensing data sources and other related information are:
http://www.itc.nl/~bakker/

3.7.5.2. Imaging spectroscopy

Spectroscopy: http://speclab.cr.usgs.gov/index.html:
Links: http://www.techexpo.com/WWW/opto-knowledge/IS_resources.html
Sensor list: http://www.geo.unizh.ch/~schaep/research/apex/is_list.html/

3.7.5.3 RADAR remote sensing

Glossary of terminology:
http://ceo1409.ceo.sai.jrc.it:8080/aladine/v1.2/tutorials/glossary/agrg.html
Principle of INSAR: http://www.sciam.com/0297issue/0297massonnet.html
Imaging radar: http://southport.jpl.nasa.gov/
Radar and radar interferometry:
http://southport.jpl.nasa.gov/scienceapps/dixon/report1.html

3.8   REFERENCES

Chiu, H.Y., and Collins, W., 1988, A spectroradiometer for airborne remote sensing. *Photogrammetric Engineering and Remote Sensing*, **44**: 507–517.

Clevers, J.G.P.W., 1999, The use of imaging spectrometry for agricultural applications, *ISPRS Journal of Photogrammetry & Remote Sensing*, **54**: 299–304.

Cloutis, E.A., 1996, Hyperspectral remote sensing: evaluation of analytical techniques. *Intenational Journal of Remote Sensing*, **17**: 2215–2242.

Cracknell, A.P., 1997, *The advanced very high resolution radiometer AVHRR*. London,Taylor & Francis.

D'Souza, G., Belward, A.S., and Malingreau, J.P., 1996, *Advances in the use of NOAA AVHRR data for land applications*. Dordrecht, Kluwer Academic Publishers.

Elachi, C., 1987, *Introduction to the physics and techniques of remote sensing*. New York, Wiley & Sons.

Goetz, A.F.H., Rowan, L.C., and Kingston, M.J., 1983, Mineral identification from orbit: initial results from the Shuttle Multispectral Infrared Radiometer. *Science*, **218**: 1020–1031.

Goetz, A.F.H., Vane, G., Solomon, J.E., and Rock, B.N., 1985, Imaging spectrometry for earth remote sensing. *Science*, **228**: 1147–1153.

Hoekman, D.H., 1990, *Radar remote sensing for applications in forestry*. PhD thesis. Wageningen, Wageningen Agricultural University

Hussin, Y.A., Bijker, W., Hoekman, D.H., Vissers, M.A.M., and Looyen, W., 1999, *Remote sensing applications for forest management*; workpackage report workpackage 6; User requirement study for remote sensing based spatial information for the sustainable management of forests. ITC, Enschede, the Netherlands, 1999.

Justice, C.O., Vermote, E., Townshend, J.R.G., Defries, R., Roy, D.P., Hall, D.K., Salomonson, V.V., Privette, J.L., Riggs, G., Strahler, A., Lucht, W., Myneni, R.B., Knyazikhin, Y., Running, S.W., Nemani, R.R., Wan, Z., Huete, A.R.,van Leeuwen, W., Wolfe, R.E., Giglio, L., Muller, J.P., Lewis, P., and Barnsley, M.J., 1998, The Moderate Resolution Imaging Spectroradiometer (MODIS): Land remote sensing for global change research. *IEEE Transactions on Geoscience and Remote Sensing*, **36**: 1228–1249.

Kahle, A.B., Palluconi, F.D., Hook, S.J., Realmuto, V.J., and Bothwell, G., 1991, The Advanced Spaceborne Thermal Emission and Reflectance Radiometer (ASTER). *International Journal of Imaging Systems Technology*, **3**: 144–156.

Kramer, H. J., 1996, *Observation of the earth and its environment: survey of missions and sensors*. Berlin, Springer-Verlag.

LaBaw, C., 1987, Airborne Imaging Spectrometer 2: The optical design, In: G. Vane, A.F.H. Goetz & D.D. Norris (Eds.), *Proceedings of the Society of Photo-optical Instrumentation Engineers*, SPIE, Vol. 834.

Massonnet, D., Feigl, K.L., Rossi, M., and Adragna, F., 1994, Radar interferometric mapping of deformation in the year after the Landers earthquake. *Nature*, **369**: 227–230.

Posselt, W., Kunkel, P., Schmidt, E., Del Bello, U., and Meynart, R., 1996, Process Research by an Imaging Space Mission. In: Fuijisada, H., Calamai, G., Sweeting, M. (Eds.), *Proceedings of the Advanced and Next-Generation Satellites II* (SPIE, Toarmina, September 1996), Vol. 2957.

Rast, M., and Bézy, J.L., 1995,The ESA Medium Resolution Imaging Spectrometer (MERIS): requirements to its mission and performance of its system. In: Curran, P.J. & Robertson, C. (Eds.), *Proceedings of the RSS'95 Remote Sensing in Action*, 11–14 September 1995, Southampton, U.K. pp. 125–132.

Van der Sanden, J.J., 1997, *Radar remote sensing to support tropical forest management*. PhD thesis. Tropenbos-Guyana Series 5. Tropenbos-Guyana Programme, Georgetown, Guyana.

Van der Meer, F., 1999, Imaging spectrometry for geological remote sensing. *Geologie & Mijnbouw*, **77**: 137–151.

Van der Meer, F., 2000, Imaging spectrometry for geological applications. In: G. Meyers, R.A. (Ed.), *Encyclopaedia of Analytical Chemistry: Applications, Theory, and Instrumentation*, (New York: Wiley), Vol. A2310, 31 pp.

Van der Meer, F., and Bakker, W., 1997, Cross Correlogram Spectral Matching (CCSM): application to surface mineralogical mapping using AVIRIS data from Cuprite, Nevada. *Remote Sensing of Environment*, **61**: 371–382.

Vane, G., and Goetz, A.F.H., 1988, Terrestrial Imaging Spectroscopy. *Remote Sensing of Environment*, **24**: 1–29.

Vane, G., and Goetz, A.F.H., 1993, Terrestrial Imaging Spectrometry: Current Status, Future Trends. *Remote Sensing of Environment*, **44**: 117–126.

Vane, G., Green, R.O., Chrien, T.G., Enmark, H.T., Hansen, E.G., and Porter, W.M., 1993, The Airborne Visible/Infrared Imaging Spectrometer (AVIRIS). *Remote Sensing of Environment*, **44**: 127–143.

4

Geographic data for environmental modelling and assessment

Bradley C. Reed, Jesslyn F. Brown and Thomas R. Loveland

ABSTRACT

Access to consistent and objective environmental data is a prerequisite for modelling. Many (global) models exist that require terrestrial environmental data. They include land-atmosphere interaction models, ecosystem process models, hydrologic models, and dynamic biosphere models; as explained in Chapter 2, these models are usually of a combined inductive-deductive construction. Modellers often face a deficit of appropriate and accurate environmental data to set the initial conditions of parameters in their models, as well as to validate model output. Until recently, there were very few observations of global-scale environmental phenomena from which to construct consistent scientific databases of vegetation, soils, and climate. A significant increase in activity to fill this void has resulted in improved databases of global topography, land cover, soils, and satellite imagery that provides information on vegetation dynamics. These data sets, governed by various use and cost policies, are available from national and international organizations. Gaining access to the appropriate data sets is one hurdle that modellers must clear; another is incorporating the geospatial data into their numerical simulation and forecasting models. Improving the communication between environmental data providers and data users is crucial for improving access to data of the appropriate scale and thematic content, refining data quality, and reducing the technical barriers to working with complex geospatial data.

4.1 INTRODUCTION

Data are the fundamental building blocks of scientific inquiry. At the beginning of the 1990s, there were few objective databases designed to support modelling, especially at a global scale. There were, of course, interpreted maps of vegetation, soils, and climate, often based on broad assumptions, but no direct, or even remote, observations of global-scale environmental phenomena. Recently there has been a significant increase of activity in building geospatial databases to support broad-scale modelling applications. These activities include modelling global topography, mapping regional- and global-scale land cover, and constructing soils databases.

Environmental assessments require data that provide a means to determine the geographic location of various land resources. In addition to describing the land characteristics, assessments often require attributes that describe various properties of the landscape, such as surface roughness, biomass, and slope. As a result, the collection of environmental baseline data involves intensive and costly efforts prior to the initiation of a project. Cost serves as an important incentive for encouraging the use of existing spatial data sets as an alternative to expensive data collection and assembly.

An International Geosphere Biosphere Programme (IGBP) report states that there are two important technical requirements that affect global environmental forecasts and assessments (IGBP 1992); (1) numerical models that account for the complex interactions and feedbacks of the Earth system and (2) validated, geographically referenced environmental data sets to determine parameters for these models. The geographically referenced data include both data for documenting and monitoring global change (such as land and sea temperature, and atmospheric concentrations of carbon dioxide and other trace gases) and data that characterize important forcing functions. The IGBP report concludes with the observation that land data are required by most IGBP core projects, and these data are a critical, but missing, element in models of global ecosystems and hydrology.

In a review of the role of land cover maps, Hall et al. (1995) describe the role of environmental data in land process models. For example, in these models, vegetation community composition is used to partition the landscape into functionally different regions. The specific land cover types represented in the database are related to the biological, thermodynamic, or chemical pathways. The seasonal dynamics of land cover are critical because of their influence on patterns of latent heat flux throughout the year, as represented by changes in turbulent exchange parameters, such as surface roughness and radiation exchange variables, such as albedo. They suggest the use of land cover data with functional categories that are directly related to properties such as energy, water, and nutrient cycling. They propose a set of functional classes for use in International Satellite Land Surface Climatology Project (ISLSCP) science initiatives but concede that a single categorization of land cover types is unlikely to meet all modelling requirements.

In Kemp's (1992) review of the role of spatial data in environmental simulation models, she identifies problems of environmental data faced by modelling groups. Although Kemp does not focus on specific data sets, she states that dealing with the range of spatial and temporal scales at which different processes are depicted in models is a formidable task. Kemp suggests that modelling has been hampered by the fact that critical environmental processes operate on many different time and spatial scales, and there may be scale thresholds at which critical processes change. This is complicated by the fact that land cover, soils, and geology are typically mapped by category, but mathematical models commonly use continuous data. As a result, the categorical information must be converted to parameter values (e.g. stomatal resistance, surface roughness, leaf area index) before being input to numerical calculations. This means that researchers must use measurements from limited field or laboratory studies, in combination with their scientific 'judgement', to set parameter values for models; that is, use a combination of induction and deduction as discussed in Chapter 2.

4.2   LAND-ATMOSPHERE INTERACTION MODELLING

Climate models incorporate a greater quantity and higher quality of terrestrial data compared to a decade ago. Both general circulation models (GCM) that are used to estimate global climates under specific conditions, and mesoscale meteorological models, which operate at regional levels, make use of land cover data to set parameters for land-atmosphere interactions. Some of the more detailed parameterizations of land-atmosphere interactions are land surface schemes coupled with climate models. Fluxes of moisture, energy, and momentum at the land-atmosphere boundary are key drivers of climate models. Land surface process models were developed to describe the effects of the environmental state on these fluxes. Results from coupled models demonstrate that the biophysical properties of vegetation and the physical properties of soils are important causal agents acting on regional and global climate (Bonan 1995).

   Dickinson (1995) provides an overview of land cover scale issues in climate modelling. Traditionally, single representations of land cover were used to describe large grid cells (e.g. 40 km^2, 1° x 1°, or even 5° x 5°). Land cover data are used to assign the various parameters that characterize the role of vegetation in the model. These parameters are given as data arrays; that is, as a number or numbers for each type of land cover. One of the serious criticisms of present treatments of land processes in climate models is their failure to include many of the most essential aspects of sub grid-scale heterogeneity (Henderson-Sellers and Pittman, 1992). The latter issue can only be solved when land cover databases become available at resolutions higher than the grid cells used in the models. As previously mentioned, validation of the data sets input to the model is critical because poor quality input data will cause inaccurate output.

4.3   ECOSYSTEMS PROCESS MODELLING

There is an increasing recognition of the relationship between atmospheric process modelling and ecosystems processes and functions, including biogeochemistry and net primary productivity. Models of biogeochemistry processes and cycles are being developed and used to assess the influence of biogeochemical cycles on the physical climate system. This, in turn, has inspired significant advances in the understanding of the cycles that are responsible for the chemical compositions of the atmosphere, oceans, and surface sediments. The IGBP Global Analysis and Integrated Modeling Task Force determined that it is essential to quantify the characteristic dynamics of these cycles and their controlling factors in order to improve the understanding of global processes. This means that many types of data are needed for model input and validation of contemporary era models (IGBP 1994).

   In biogeochemical modelling, it is important to include land cover attributes describing community composition or vegetation types so that physiological differences in assimilation rates, carbon allocation, and nutrient use efficiency, which influence CO_2 uptake during photosynthesis, can be modelled (Bonan 1995). Matson and Ojima (1990) stress a growing requirement for land cover data in atmospheric chemistry studies because of a need for data that can be used for the synthesis of global N_2O, CH_4, and CO_2 fluxes. Models that include detailed canopy physiology, energy exchange, and microbial processes have been developed (e.g., the CENTURY model of Parton *et al.* 1988; BIOME-BGC developed by Hunt *et al.* 1996). These models have been used to evaluate the equilibrium response of ecosystems to doubled atmospheric CO_2 and associated climate change.

Issues concerning appropriate scale of study are also apparent in this type of modelling. Kittel *et al.* (1996) suggest that the accurate representation of the spatial distribution of driving variables and boundary conditions is necessary for simulations of regional to global ecological dynamics. However, developing such representations is difficult due to a lack of regional data sets and because of the coarse resolution needed to cover large domains. When land cover and soils are heterogeneous with respect to the model simulation grid, grid averages may not adequately represent existing conditions. As a result, data must have sufficient spatial detail for the statistical treatment of spatial heterogeneity and spatial coherence among variables that have complex relationships to ecosystem processes. The ability to incorporate sub grid information (i.e., estimates of the percentages of different land cover types within a grid cell) instead of using finer grid cells is necessary to keep the already high computational requirements of large-area simulations at a manageable level. This can only be accomplished, they state, with (1) the development of physically consistent model input data sets, (2) the efficient transfer of data between geographic information systems (GIS) and applications, and (3) the analysis of model results in both geographic and temporal contexts.

4.4 HYDROLOGIC MODELLING

Elevation, elevation derivatives (e.g. slope, aspect, drainage flow direction) and land cover data are required for modelling the physical processes of the hydrologic cycle (Steyaert *et al.*, 1994). However, land cover is frequently treated as uniform throughout a watershed. Vegetation interacts with the atmosphere by its influence on the return of water back to the atmosphere through evaporation and transpiration, thereby exerting a significant impact on the hydrologic cycle by influencing the exchange of energy, water, carbon, and other substances at the land surface-atmosphere interface. The IGBP Biological Aspects of the Hydrologic Cycle (BAHC) core project aims to serve as a catalyst for the improving future hydrologic models. BAHC plans stress the need to explicitly assess and represent the relevant topographic, vegetation, soil, and geologic parameters that control soil moisture and surface and subsurface hydrologic conditions in order to improve the results of hydrologic simulations (IGBP 1993).

4.5 DYNAMIC BIOSPHERE MODELLING

Advances in equilibrium terrestrial modelling have led to an evolution towards a dynamic biosphere model. Tian *et al.* (1998) describe the goals and challenges associated with modelling the terrestrial biosphere and accounting for its truly dynamic character. This 'new generation' model would ideally demonstrate ecosystem dynamics caused by natural or anthropogenic factors, in addition to the changing interactions and feedback among the cycles of energy, water, and biogeochemical elements. Again, data are lacking to help make models more mechanistic and to support model validation. Tian *et al.* (1998) call for measurements of whole ecosystem behavior, such as carbon and water fluxes, in addition to vegetation dynamics and land use change in order to make progress towards a dynamic biosphere model.

4.6 DATA ACCESS

Identifying and securing data that support environmental modelling and assessments is often a frustrating process. Data are commonly produced with a specific customer or application in mind and are typically not easily adapted to the needs of other applications. Often one must deal with vexing problems regarding data format, resolution, and content. In addition, simply finding the best data for an application is time consuming and difficult. For regional-scale applications, it is especially difficult to sort through all of the possibilities for data, many of which are under publicized, incompletely documented, and not fully supported. Consequently, data sets that become widely used are often 'advertised' by word-of-mouth and subsequent exchange of data among modellers. Data for developing countries are especially difficult to obtain as data sources are often outdated, unreliable, or simply unavailable. For many developing countries, the only recourse is to resort to coarser data at regional, continental, or global scales.

The number of primary data sources is limited for continental- and global-scale modelling applications. Although by no means complete, a good start towards acquiring global- or coarse-scale data at little or no cost is through one of the following sources:

*   US Geological Survey's (USGS) Earth Resources Observation Systems (EROS) Data Center (edcwww.cr.usgs.gov)
*   NASA Goddard Space Flight Center (GSFC) Distributed Active Archive Center (daac.gsfc.nasa.gov)
*   United Nations Environment Programme Global Resources Information Database (UNEP-GRID) (grid2.cr.usgs.gov)
*   The European Commission Joint Research Centre, Space Applications Institute (JRC, SAI) (www.sai.jrc.it).

Data access policies of the many data providers vary widely but, encouragingly, are starting to become more liberal. For example, UNEP-GRID has a policy that allows U.N. organizations, inter-governmental organizations, scientific and academic organizations, and non-governmental organizations to request data. However, data requests from private firms and individuals will not be filled except in exceptional cases.

The US Government policy calls for cost recovery, with data prices based on the cost of filling user requests rather than on the value of the data. The USGS, one of the major spatial data providers in the US, follows this policy. The USGS is also moving toward a goal of making digital data available at no cost via electronic transfer. However, full no-cost access to electronic data will not be feasible until bandwidth increases for all providers and users

Recently JRC-SAI and the European Umbrella Organisation for Geographic Information (EUROGI) sponsored a joint workshop to address the issue of geographic data policy in the European Union (Craglia *et al.*, 2000). Policies on geographic data vary widely in Europe. With respect to pricing, there is often a distinction between 'essential' data, such as ownership, administrative boundaries and roads, and 'non-essential' data that are value-added. Essential data are often accessible free of charge, while non-essential data may have a charge. Where there are charges for data, the policy usually states something to the effect that price should not deter the use of data. In reality, it is difficult to sort through all of the various data access issues, and this, unfortunately, discourages data usage. With the rapid developments in computing and the Internet, many mapping agencies are being forced to re-examine data policies. Among actions resulting from the workshop, the European Commission seems ready to establish a framework for developing a more cohesive geographic information policy.

4.7 GLOBAL DATABASES

The amount of geographic data available for environmental applications has rapidly increased over the past two decades. The following section describes widely available land-related data that are commonly applied in broad-scale environmental modelling using GIS and remote sensing.

4.7.1 Multiple-theme global databases

In recent years, ISLSCP has performed a lead role in the Global Energy and Water Cycle Experiment (GEWEX). Its primary responsibilities concern land-atmosphere interactions, with special focus on process modelling, data retrieval algorithms, field experiment design, and development of global data sets to supply modellers with data for land-biosphere-atmosphere models. The data sets provide fields for model initialization and forcing functions. In 1994, the ISLSCP Initiative I data sets were produced and published.

The data, distributed both on CD-ROM and online, are organized into five groups:

• Vegetation
• Hydrology and Soils
• Snow, Ice, and Oceans
• Radiation and Clouds
• Near-Surface Meteorology.

The ISLCSP team gathered data from various sources and processed each data layer for spatial and temporal consistency. Generally, the Initiative I global data layers are distributed on a 1° by 1° grid and cover a 24-month period (1987–1988) at monthly time steps. Building on the experience of the first initiative, a project to produce the ISLSCP Initiative II data collection project was launched in 1999 with funding from NASA's Hydrology Program. The project objectives include producing and validating an expanded series of global data sets. Meterological data are produced at a 1° x 1° spatial resolution; topography, soils, and vegetation layers will be produced at finer scales (half- and quarter-degree). These data layers make up a comprehensive set covering a 10-year period from 1986 to 1995. Prototype data are scheduled for release in late 2000, with the publication of final data sets scheduled for 2002. More information and ftp access to data layers can be found on the Internet (http://islscp2.gsfc.nasa.gov/).

Another multi-theme database is the Digital Chart of the World. This data set includes a global base map with many features and attributes including administrative boundaries, coastlines, hydrologic features, and urban features. The database is distributed from the Environmental Systems Research Institute, Inc (ESRI). This is a private company in the US that also distributes geospatial software. This data set is distributed on CD-ROM media for a cost. It can be accessed from the ESRI web page (at http://www.esri.com).

Other multiple-theme global databases include the Global Map 2000 project. The Japan Ministry of Construction first proposed the concept of the Global Map 2000 in 1992, and in 1994 the Geographical Survey Institute of Japan proposed the first specifications for the project. Global Map 2000 currently involves the participation of over 70 countries. The main objective of the project is to bring mapping agencies from across the world together to develop and provide easy and open access to global digital geographic information at a 1:1,000,000 scale, an appropriate scale to ratify and monitor global agreements and conventions for environmental protection, and to encourage economic growth within the context of sustainable development. To be viable, the initial project will use existing global data whenever possible and will also seek to improve the reliability and accuracy of these data sets. The upgrading of existing data is expected to involve editing or replacing parts of it with data from other sources. For more information, see (http://www1.gsi-mc.go.jp/iscgm-sec/index.html).

4.7.2 Heritage global land cover databases

In the 1980s, efforts were undertaken to develop very coarse resolution databases (approximately 1° x 1°) that could be used by researchers working with GCMs. The global databases combined existing maps with different cover classes into coarse-scale mosaics of categorical data. The source maps were sometimes out of date and usually required a translation in class names between the map classification and the desired classification scheme. The procedures to create appropriate land surface data for modelling included extensive reclassification or aggregation of pre-existing vegetation information.

The Olson and Watts (1982) database, generally referred to as the Olson Global Ecosystems map, originally consisted of 49 categories. These categories represented natural, anthropogenic, or natural/modified mosaics of land cover. The Olson database, with a spatial resolution of 0.5° x 0.5°, has undergone continual updating and improvement. The current Olson Global Ecosystems legend now consists of 96 land cover classes. The Matthews (1983) global database consists of two data layers; the primary layer is a 32-category map of potential natural vegetation with a grid resolution of 1° x 1° and the secondary layer includes estimates of the percentage of cultivated land for each grid cell. The two layers can be combined to produce a map that approximates actual land cover. Like the Olson map, the Matthews map was compiled from an exhaustive review of hundreds of source materials, including maps, texts, atlases, and statistical summaries. A third global database is the Wilson and Henderson-Sellers global land cover database (1985). This database has a spatial resolution of 1° x 1° and contains 53 categories of land cover. Each grid cell includes a primary and secondary land cover attribute. The land cover categories consist of physiognomic elements, including canopy density, and have inferences to land use activities.

4.7.3 Global land cover from satellite data

One of the driving forces improving global land data sets was the IGBP-Data and Information System's (DIS) promotion of the development of a 1 km global land data product from the National Oceanic and Atmospheric Administration's advanced very high resolution radiometer (NOAA-AVHRR) (Eidenshink and Faundeen, 1994). The development of the data set, begun in 1992, was sponsored by the USGS, NASA, NOAA, and the European Space Agency (ESA). This data set provides satellite observations processed in a consistent manner into 10-day composite data.

The Land Cover Working Group (LCWG) of the IGBP-DIS was responsible for designing and fostering the development of an improved global land cover data set. Through a series of international workshops, the LCWG finalized a strategy and definition for a global land cover product (DISCover) based on 1 km AVHRR data (Belward 1996). The USGS EROS Data Center (EDC), in partnership with national and international agencies and universities, developed a global land cover characteristics database described in detail by Loveland *et al.* (2000). The database was constructed through the unsupervised classification of continental-scale, time-series normalized difference vegetation index (NDVI) data with extensive post-classification stratification. The global database contains 961 distinct seasonal land cover regions that are translated into a number of more general, predefined classification schemes, such as the USGS Anderson System, IGBP, Olson Ecosystems, and those required by the Biosphere Atmosphere Transfer Scheme (BATS) and Simple Biosphere (SiB) models. The database layers, formatted as raster grid images, are distributed free of charge at (http://edcdaac.usgs.gov/glcc/glcc.html).

The University of Maryland also produced a global land cover product using the same base AVHRR data, but the data were classified using a classification tree approach (Hansen *et al.* 2000). Several metrics derived from temporal AVHRR data, such as minimum red reflectance, peak NDVI, and minimum channel-three brightness temperatures, were used to predict class memberships. The resulting product contains 14 land cover classes. These data are available through the University of Maryland Laboratory for Global Remote Sensing Studies (http://www.geog.umd.edu/landcover/global-cover.html).

The EDC and University of Maryland global land cover products represent significant strides toward resolving the land cover information needs of environmental modellers. These data sets provide consistent global products that can be used to estimate parameters for land-atmosphere interactions with sufficient spatial resolution to describe subgrid cell heterogeneity for most models. The primary weaknesses of the products are those imposed by the limitations of the AVHRR satellite sensor. Many of these weaknesses will be addressed in current efforts at land cover characterization using data from the SPOT Vegetation sensor and the Moderate Resolution Imaging Spectroradiometer (MODIS).

4.7.4 Topographic data

One common input to a wide variety of terrestrial and land-atmosphere interaction models is topographic or digital elevation model (DEM) data. The recently developed global model, GTOPO30, reflects a significant improvement in the quality and resolution of DEMs over previous models (Gesch *et al.* 1999). Elevations are spaced at 30-arc-second intervals (about 1 km) with a reported overall vertical accuracy of 70 m.

A team from the USGS EDC assembled topographic data from eight different sources to derive elevation values for GTOPO30. The primary source was the Digital Terrain Elevation Data (DTED) product of the National Imagery and Mapping Agency. GTOPO30 has no use or redistribution restrictions and is available at low or no charge, depending on the user's choice of access. The global DEM is distributed through the EROS Data Center Distributed Active Archive Center. Users may either download the data electronically (the global dataset is distributed in tiles) or order the data on CD-ROM for a nominal fee via the Internet: (http://edcdaac.usgs.gov/gtopo30/gtopo30.html).

The EDC is also creating a HYDRO1k data set for use in hydrologic and morphometric studies. Development of the HYDRO1k database was made possible by the completion of the 30-arc-second DEM (GTOPO30) at the EDC in 1996 (Gesch *et al.* 1999). A goal of the HYDRO1k project is the systematic development of hydrologically sound derivative products for all the land masses of the globe (with the exception of Antarctica and Greenland). Additionally, topologically sound representations of the Earth's rivers and basins are being developed, which can be used for comparison among rivers at corresponding scales.

The HYDRO1k data sets are being developed on a continental basis and provide a suite of six raster and two vector data sets. The raster data sets are the hydrologically correct DEM, derived flow directions, flow accumulations, slope, aspect, and a compound topographic (wetness) index. The derived streamlines and basins are distributed as vector data sets. The HYDRO1k data sets are available over the Internet at (http://edcdaac.usgs.gov/gtopo30/hydro/index.html).

4.7.5 Soils data

The FAO has produced a global soil texture data set known as GLOBTEX (FAO 1970). This digital database for the Soil Map of the World represents the most comprehensive available global soils data. The source maps were compiled by the FAO and published by UNESCO in 10 volumes between 1970 and 1978. The map references 106 FAO soil types and includes the texture and slope of the dominant soil in each area. Texture is defined in three classes (coarse, medium, and fine) and slope is defined in three classes (0–8 percent, 8–30 percent, and >30 percent slopes).

Several attempts have been made to generalize the FAO data for use in climate models (Zobler 1987; Wilson and Henderson-Sellers 1985). They encoded the FAO maps on 1-degree or half-degree cells and considered only the characteristics of the dominant soil of the cell. In 1984, the FAO map was digitized by the ESRI. The resulting database includes the full spatial detail contained on the original published maps and the associated soils (approximately 20 per cent by area) and included soils (approximately 10 per cent by area) within the map units.

A digital version of the database is available through the Web page for UNEP-GRID at (http://grid2.cr.usgs.gov/data/fts.html).

4.7.6 Global population

Population data sets provide significant information concerning the human-factor in modelling. One data set, the Gridded Population of the World, is distributed by CIESEN (http://www.ciesen.org). The data set includes population for multiple years (1990–1995) at a 2.5 minute grid cell resolution. Another world-wide database of ambient population for the year 1998 has a 30-second resolution. It was made using census counts and then deriving probability coefficients based on a number of information layers (roads, slope, land cover, and night-time lights satellite data). The data set, known as LandScan, was created at Oak Ridge National Laboratory, US Department of Energy and is available from (http://www.ornl.gov/gist/projects/LandScan/landscan.doc).

4.7.7 Satellite data

Although most environmental modelling applications require processed information (e.g., land cover, elevation, or soils), some models are being adapted to directly enter data collected from remote sensing satellites. The choice of which satellite data to use becomes a question of spatial resolution, spectral characteristics of the satellite sensor, the frequency of coverage, and the areal coverage or extent (swath size). Table 3.1 provides a summary of the more commonly used satellite sensors. These sensors vary in swath width from 12 km to 2400 km and in spatial resolution from 1 m to over 1 km. Generally, the cost of these products is inversely proportional to the spatial resolution. The resolution of satellite data used in models must satisfy information requirements through spatial discrimination of land features, providing flexible, subgrid cell parameterization, with a data volume that can be accommodated in computationally complex models. Data sets with fine spatial detail have costly data storage requirements that increase the time required for modelling exercises. Thus, a balance must be found between spatial resolution and associated data volume efficiency.

Most models require data covering large land areas, which until recently meant the use of AVHRR data. These data have been analyzed, scrutinized, and criticized for nearly two decades. There is little doubt that, even with the shortcomings of AVHRR data, studies using this sensor have served as a strong bridge between data providers and environmental modellers. The encouraging results gained from AVHRR studies in the field of land cover characteristics (e.g., Loveland *et al.* 2000; Hansen *et al.* 2000) and the derivation of biophysical parameters from land cover have helped to drive requirements for improved large-area sensors, such as SPOT4 Vegetation and MODIS. MODIS, in particular, is specifically designed to provide new, innovative land products that are especially designed to support modelling applications (Justice 1998). In addition to providing satellite data, the MODIS Land Group is providing higher-level data products that are specifically designed to support the needs of global to regional monitoring, modelling, and assessment, such as leaf area index, net primary production, albedo, and others. For the latest information on MODIS land products, see: (http://modis-land.gsfc.nasa.gov).

4.8   SUB-GLOBAL SCALE DATABASES

In addition to the global efforts described in previous sections, there are numerous environmental data sets that cover large but sub-global regions. Presenting a comprehensive treatment of these databases is beyond the scope of this chapter. However, this section will identify some sources of sub-global data sets and provide examples.

Regional or national datasets cover as many data themes as their global counter parts (i.e. land cover, topography, climate, economic, population), and have varied sources, costs, and distribution policies. These types of data sets are often even more difficult to locate than the global data mentioned above. The advantages they offer may include greater spatial detail, better local and regional consistency, and improved accuracy.

4.8.1 Regional land cover mapping

There are several examples of relatively current regional land cover databases. The National Land Cover Database (NLCD) covering the conterminous US is produced by the USGS EDC and is based on 1992-era Landsat TM imagery. It is available at no charge from (http://landcover.usgs.gov/nationallandcover.html). The NLCD maps land cover into a 21-class legend. The CORINE (coordination of information on the environment) database, produced by the European Environment Agency and JRC, is based on both SPOT and Landsat data and covers most of the European Union. This database is produced from 1990-era data and is available through (http://natlan.eea.eu.int/datasets.html). Both the NLCD and CORINE databases will be updated based on year 2000 satellite data, which will provide valuable information on land cover change during the 1990s.

Another project in Europe, Pan-European Land Use and Land Cover Monitoring (PELCOM) had the objective of creating 1 km resolution land use and land cover data from AVHRR data that could be updated frequently. The project was led from the Netherlands (ALTERRA/Green World Research), and had international project partners from Europe. This land cover data set is available at http://137.224.135.82/cgi/geodesk/geodata/pelcom.html.

The Africover project, sponsored by the Food and Agriculture Organization (FAO) of the United Nations, has a goal of establishing a digital database on land cover for all of Africa at a 1:200,000 scale (1:100,000 scale for small countries). The implementation of the project in East Africa is currently best developed (http://www.africover.org). For a more extensive review of activities in global and regional land cover mapping, see DeFries and Belward (2000).

4.8.2 Topographic databases

There are a number of national-level topographic data sets available. Many of them are accessible from the appropriate government mapping agency. In Australia, for example, the Australian Surveying and Land Information Group (AUSLIG) is responsible for distributing a number of different topographic data sets. They include gridded digital elevation data at several different resolutions (for example, 18-second, nine-second, and three-second resolution data) and point elevation data sets. Data sets may be accessed from their site (http://www.auslig.gov/products/).

In the US, the National Elevation Dataset (NED) has been developed by the USGS by merging high-resolution elevation data into a seamless raster format. The data set has a resolution of one arc-second resolution for the conterminous US, Hawaii, and Puerto Rico and two arc-seconds for Alaska. NED will be updated periodically to incorporate the highest resolution and quality data possible. In areas where 7.5 minute DEM data are available, those data are used to generate NED). Where they are not yet available, coarser resolution DEM data (for example, 30 minute or one degree DEM) products are used to generate NED. Users may order their area of interest interactively, and there is a cost for the data, online at http://edcnts12.cr.usgs.gov/ned/.

4.8.3 Administrative and census data

Administrative data, including political boundaries, administrative features, and place names, are frequently distributed by national mapping agencies. In Britain, the Ordnance Survey produces 1:10,000 scale raster data in tiff format. It is sold in 5 by 5 km tiles. The basic product includes information on place names, houses, streets, rivers, etc. Access to the Ordnance Survey of Britain can be found at http://www.ordnancesurvey.co.uk/. Similarly, in Ireland, the Ordnance Survey Ireland has a mandate to provide three types of map services to Ireland. These are detailed large scale maps at 1:1,000 scale in urban locales, rural areas covered by 1:2,500 and 1:5,000 scale maps, and tourist maps at 1:50,000 scale. Access to all of these products is through http://www.irlgov.ie/osi/.

Much of the US Census Bureau data (http://www.census.gov/) is inherently statistical. However, they distribute a number of geographic products with versions available reflecting the year 2000 census data. The Census 2000 Tiger/Line Files is the public version of the US Census Bureau's digital database of geographic features for the US This product includes Census 2000 tabulation block numbers, address ranges, and governmental unit boundaries. They also distribute a number of maps depicting the geographic units used for census data collection, including census tract outlines, census blocks, voting districts, and state legislative district outlines.

4.8.4 Data clearinghouses

Some organizations have missions to provide data for applications that require regional environmental data. Data clearinghouses, with entry points on the Internet that provide access to data and metadata, are becoming more common as technology has developed. The following examples outline several different types of clearinghouses.

The Africa Data Dissemination Service (ADDS) is managed by the US Agency for International Development (USAID) in collaboration with the USGS to support the Famine Early Warning System Network. The project goals are to identify areas that might experience food supply problems due to famine, flood, or other environmental conditions. The ADDS includes data from 17 sub-Saharan African countries. Many multi-theme data sets are distributed from the ADDS (located at http://edcintl.cr.usgs.gov/adds/adds.html). These include satellite data, administrative boundaries, digital elevation models, hydrology, roads, and vegetation. Tabular statistics by country are also available on agriculture and precipitation.

As mentioned earlier, the UNEP-GRID provides data to assist the international community and individual nations to make sound decisions on environmental planning and resource management. The GRID network consists of several nodes. The Geneva node (http://www.grid.unep.ch/) supports access to over 200 data sets, both global and sub-global in geographic coverage. Multiple data themes are included, such as administrative boundaries, atmosphere, biodiversity, oceans, climate, and soils. The North American GRID node (http://grid2.cr.usgs.gov/) node also distributes data sets on population and administrative boundaries.

The Australian Department of the Environment and Heritage supports its own data clearinghouse, the Australia Spatial Data Directory (at http://www.environment.gov.au/search/databases.html). Data sets from a variety of themes, such as air, biodiversity, land conservation, and industry, are all linked from this location. The site also supports an online map-making function which will add layers interactively depicting features such as hydrology, infrastructure, conservation areas, and species distributions.

The Inter-American Geospatial Data Network (IGDN), formed by USAID in partnership with the USGS, promotes information infrastructure (especially Internet capabilities) in the Western Hemisphere for electronic access to information on geospatial data. This site (at http://edcsnw3.cr.usgs.gov/igdn/in-dex.html) does not house data sets, but does provide links and searching capabilities to geospatial data providers throughout the Western Hemisphere.

4.9   THE ROLE OF THE END-USER IN THE USGS GLOBAL LAND COVER CHARACTERIZATION PROJECT

Improving communication between data users (e.g., modelers) and the producers of land surface environmental data has been a key component to improving terrestrial data sets. If a data set is to meet the diverse data requirements of a broad environmental modelling community, then avenues for positive communication flow must be opened. One example of incorporating user input is the USGS global land cover characterization project.

User feedback (that is, the comments, suggestions, constructive criticism, and requests from the users of the global land cover database) played a vital role in both setting the requirements for the global land cover effort and assessing the quality and utility of the database through the peer review process (Brown *et al.* 1999). The IGBP global land cover strategy explicitly called for a validation protocol to assess the accuracy of the DISCover product (Belward 1996). This process consisted of three primary activities: (1) peer review of the preliminary continental databases, (2) comparisons with other land cover data sets, and (3) a formal, statistically sound, accuracy assessment. The USGS mapping team used peer review not only to identify and correct classification problems, but to gain feedback on a wider range of database issues, including data utility in applications, suggestions as to data set improvements, and insights into the technical challenges faced by users working with the data. Users provided many suggestions; frequently identifying areas with land cover label problems. The comments, when collaborated using other evidence, were a critical resource for refining the global land cover characteristics database, leading to the generation of a new version, released in 2000. Approximately 10 per cent of the land area has been revised since the first release, largely based on user feedback.   In addition, improvements in documentation, alternative projections, and additional classification schemes have all been added to the database because of open communication with users.

The accuracy assessment was performed in accordance with the specifications developed by the IGBP Validation Working Group and approved by the Land Cover Working Group. The approach was a stratified random sample of DISCover and the use of higher spatial resolution satellite imagery (either Landsat TM or SPOT) to test class accuracy (Scepan 1999). After consideration of both time and personnel required to execute the accuracy assessment, a random sample was taken within each of the DISCover classes (excluding snow and ice and water) with a target accuracy of 85 per cent and a confidence level of 95 per cent. A manageable sample size of 25 for each class, based on the expected accuracy and confidence level results in an interval with a range of +/- 0.143 (Scepan 1999).

The validation samples were aggregated into 13 regions and several expert image interpreters (EII) were enlisted for image interpretation for each of the regions. The EIIs assembled for validation workshops and three EIIS were assigned to validate each sample using the high resolution imagery. The VWG protocol specified that at least two of three EIIs must agree on the land cover type before a given DISCover sample was accepted as correct. If all three disagreed and identified three different land cover types for a sample, the DISCover classification was considered to be in error. The average class accuracy was 59.4 per cent with single class accuracies ranging between 40 and 100 per cent. If the accuracies are weighted according to class size, the area weighted accuracy rises to 66.9 per cent (Scepan 1999). For more detail on the validation procedure and related studies see the special issue of Photogrammetric Engineering and Remote Sensing: Global Land Cover Data Set Validation, September 1999.

4.10 SUMMARY

A variety of models have a growing need for improved environmental data sets to better approach biophysical parameterization. Significant advances have recently been made in the development of large-area databases to support such models and environmental assessments. Consistency in database construction, as well as the scale and areal coverage of databases, is improving. This allows issues such as subgrid cell heterogeneity to be better addressed. Rapid improvements in technology, computing, and the Internet, along with the decrease in the cost of disk storage, have changed the arena of land surface mapping. Several large area databases are now available free of charge by means of the Internet. Although recent developments have resulted in improved environmental data, there are still serious impediments to using the data owing to proprietary data formats, technology limitations (i.e. bandwidth), inconsistent data policies, and the inherent complexity of spatial data (e.g. disparate data resolutions, map projections, etc.). It appears that governmental data distribution policy shifts will continue to encourage the availability of high-quality, low-cost data. Even though there is a trend towards improving data, it is necessary for environmental data providers and environmental modelling and assessment researchers to continue to keep lines of communication and collaboration open to improve our understanding of the Earth's environment.

4.11 REFERENCES

Belward, A.S. (ed.), 1996. *The IGBP-DIS Global 1-Km Land Cover Data Set (DISCOVER): Proposal and Implementation Plans.* IGBP-DIS Working Paper No. 13, IGBP-DIS Office, Meteo-France, 42 Av. G. Coriolis, 31057 Toulouse, France, 61 pp.

Bonan, G.B., 1995, Land-Atmosphere Interactions for Climate System Models: Coupling Biophysical, Biogeochemical and Ecosystem Dynamical Processes. *Remote Sensing of Environment*, **51**, 57–73.

Brown, J.F., Loveland T.R., Ohlen D.O. and Zhu Z., 1999, The global land-cover characteristics database: the users' perspective. *Photogrammetric Engineering & Remote Sensing*, **65**, 1069–1074.

Craglia, M., Annoni A. and Masser I. (eds), 2000, In: *Proceedings of the EUROGIEC Data Policy Workshop: Geographic Information Policies in Europe: National and Regional Perspectives*, European Communities, 42 pp.

DeFries, R.S. and Belward A.S. (eds), 2000, Global and regional land cover characterization from remotely sensed data. *Special Issue: International Journal of Remote Sensing*, **21**, 1081–1560.

Dickinson, R.E., 1995, Land Processes in Climate Models. *Remote Sensing of Environment*, **51**, 27–38.

Dobson, J.E., Bright, E.A, P.R., Durfee, R.C. and Worley, B.A., 2000, Landscan: A global population database for estimating populations at risk. *Photogrammetric Engineering and Remote Sensing*, **66**, 849–857.

Eidenshink, J.C. and Faundeen, J.L., 1994, The 1 km AVHRR global land data set: first stages in implementation. *International Journal of Remote Sensing*, **15**, 3443–3462.

Food and Agriculture Organization of the United Nations, 1970–1978, *Soil Map of the World*, Scale 1:5,000,000. Volumes I-X. United Nations Educational Scientific and Cultural Organization, Paris.

Gesch, D.B., Verdin, K.L. and Greenlee, S.K., 1999, New land surface digital elevation models cover the earth. *EOS, Transactions, American Geophysical Union*, **80**, 69–70.

Hall, F.G., Townshend, J.R. and Engman, E.T., 1995, Status of Remote Sensing Algorithms for Estimation of Land Surface Parameters. *Remote Sensing of Environment*, **51**, 138–156.

Hansen, M.C., Defries, R.S., Townshend, J.R.G. and Sohlbert, R., 2000, Global land cover classification at 1 km spatial resolution using a classification tree approach. *International Journal of Remote Sensing*, **21**, 1331–1364.

Henderson-Sellers, A. and Pitman, A.J., 1992, Land-Surface Schemes for Future Climate Models: Specification, Aggregation and Heterogeneity. *Journal of Geophysical Research*, **97**, 2687–2696.

Hunt, E.R. Jr., Piper, S.C., Nemani, R., Keeling, C.D., Otto, R.D. and Running, S.W., 1996, Global Net Carbon Exchange and Intra-Annual Atmospheric CO_2 Concentrations Predicted by an Ecosystems Process Model and Three-Dimensional Atmospheric Transport Model. *Global Biogeochemical Cycles*, **10**, 431–456.

International Geosphere Biosphere Programme, 1992, *Improved Global Data for Land Applications*, J.R.G. Townshend, (ed.) IGBP Global Change Report No. 20, International Geosphere Biosphere Programme, Stockholm, Sweden, 87 pp.

International Geosphere Biosphere Programme, 1993, *Biospheric Aspects of the Hydrological Cycle (BAHC): The Operational Plan*. IGBP Global Change Report No. 27. International Geosphere Biosphere Programme, Stockholm, Sweden, 103 pp.

International Geosphere Biosphere Programme, 1994, *IGBP Global Modeling and Data Activities 1994–1998*. IGBP Global Change Report No. 30, International Geosphere Biosphere Programme, Stockholm, Sweden, 87 pp.

Justice, C.O. (ed.), 1998, The moderate-resolution imaging spectroradiometer (MODIS); land remote sensing for global change research. *TRANS IEEE Geoscience and Remote Sensing*, **36**, no. 4.

Kemp, K.K., 1992, Spatial Models for Environmental Modeling with GIS. In: *Proceedings: 5th International Symposium on Spatial Data Handling*, v. 2. International Geographic Union, Charleston, SC 524–533.

Kittel, T.G.F., Ojima, D.S., Schimel, D., McKeown, D., Bromberg, R., Painter, J., Rosenbloom, T. and Parton, W., 1996, Model GIS Integration and Data Set Development to Assess Terrestrial Ecosystem Vulnerability to Climate Change. In: *GIS and Environmental Modeling: Progress and Research Issues*, M.F. Goodchild, L. Steyart, B. Parks, (eds) GIS World, Inc., Fort Collins, CO: 293–297.

Loveland, T.R., Reed, B.C., Brown, J.F., Ohlen, D.O., Zhu, Z., Yang, L. and Merchant, J.W., 2000, Development of a global land cover characteristics database and IGBP DISCover from 1 km AVHRR data. *International Journal of Remote Sensing*, **21**, 1303–1330.

Matthews, E., 1983, Global Vegetation and Land Use: New High Resolution Data Bases for Limited Studies. *Journal of Climatology and Applied Meteorology*, **22**, 474–487.

Matson, P.A. and Ojima, D.S., 1990, *Terrestrial Biosphere Exchange with Global Atmospheric Chemistry: Terrestrial Biosphere Perspective of the IGAC Project*. IGBP Global Change Report 13, International Geosphere Biosphere Programme, Stockholm, Sweden, 103 pp.

Olson, J.S. and Watts, J.A., 1982, *Major World Ecosystem Complex Map*. Oak Ridge, TN: Oak Ridge National Laboratory.

Parton, W.J., Stewart, J.W.B. and Cole, C.V., 1988, Dynamics of C, N, P and S in Grassland Soils: A Model. *Biogeochemistry*, **5**, 109–131.

Scepan, J., 1999, Thematic validation of high-resolution global land-cover data sets. *Photogrammetric Engineering & Remote Sensing*, **65**, 1051–1060.

Steyart, L.T., Loveland, T.R., Brown, J.F. and Reed, B.C., 1994, Integration of Environmental Simulation Models with Satellite Remote Sensing and Geographic Information Systems Technologies: Case Studies. In: *Proceedings: Pecora 12 Symposium on Land Information from Space-Based Systems*, American Society of Photogrammetry and Remote Sensing, Bethesda, MD, 407–417.

Tian, H., Hall, C.A.S. and Qi, Y., 1998, Modeling Primary Productivity of the Terrestrial Biosphere in Changing Environments: Toward a Dynamic Biosphere Model. *Critical Reviews in Plant Sciences*, **15**, 541–557.

Wilson, M.F. and Henderson-Sellers, A., 1985, A Global Archive of Land Cover and Soils Data for Use in General Circulation Models. *Journal of Climatology*, **5**, 119–143.

Zobler, L., 1987, A world soil hydrology file for global climate modelling. p. I-229-244. In: *International Geographic Information Systems Symposium: The Research Agenda*, November 15-18, 1987, Arlington, Virginia, Proceedings, Vol. 1, Association of American Geographers. National Aeronautics and Space Administration, Washington, D.C.

Any use of trade, product, or firm names is for descriptive purposes only and does not imply endorsement by the US Government.

The biosphere: a global perspective

S.O. Los, C.J. Tucker, A. Anyamba, M. Cherlet, G.J. Collatz,
L. Giglio, F.G. Hall and J.A. Kendall

5.1 INTRODUCTION

Satellite monitoring of land surface vegetation at regional and global scales is the topic of this chapter. The chapter provides an overview of developments during the past 20 years, highlights several applications, and discusses pitfalls in the analysis of multi-year satellite data.

Monitoring of land surface vegetation from satellites at regional and global scales is most commonly used to detect year-to-year variations in vegetation. Variations in vegetation can provide us with estimates of interannual variation in crop yield, movements of desert boundaries, and indications of environmental conditions that are associated with the outbreak of pests and diseases. Some of the variations in vegetation are related to quasi-periodic climate oscillations (e.g. El Niño), others are related to the conversion of land, e.g., from a tropical forest into cropland. All applications require data that are comparable in space and time.

Some of the vegetation processes we study with satellite data are of immediate importance; e.g., detection of failed crops or of environmental conditions that are associated with outbreaks of pests and diseases is essential for policy makers to take appropriate action. Other vegetation processes may have importance for longer time-scales, e.g., long-term monitoring of responses of vegetation to variations in climate could help us understand feedbacks between climate and vegetation and improve projections of the impacts of climate and environmental change.

In this chapter several applications related to global and regional monitoring of vegetation are discussed. Most of these applications are based on NOAA Advanced Very High Resolution Radiometer (AVHRR) or Landsat data. The NOAA and Landsat satellites have been in operation for several decades and from the long records collected by these instruments we can obtain meaningful estimates of variations in land surface vegetation during the past two decades. No other globally comprehensive data series provides us with this unique capability to study vegetation over a comparable period at global scales.

Vegetation monitoring at global and regional scales began with the launch of NOAA-7 in 1981. The AVHRR aboard NOAA-7 collected data from the entire globe at least once during daytime and this provided us with the opportunity to obtain frequent updates of vegetation at regional and global scales. The spatial resolution of the AVHRR is much less than of Landsat, hence the possibility to frequently monitor vegetation conditions comes with a loss of spatial detail.

Both the AVHRR and earlier Landsat scanners were first generation instruments and have flaws in their design. As a result, data are often not spatially and temporally consistent, even though this is a first requirement for meaningful detection of year-to-year variations in vegetation. Often, this requirement of spatial and temporal consistency of vegetation data was an afterthought in the design of previous satellite remote sensing systems. This chapter will show examples of ways to deal with these issues and to extract meaningful information from corrected data. New satellite missions such as SPOT-VEGETATION, TERRA, and Landsat 7 that have been launched recently are designed specifically for vegetation monitoring and are expected to have fewer data problems as a result.

5.2 HISTORIC OVERVIEW

Remote sensing of vegetation at regional scales began with the launch of the Landsat satellite in the mid-seventies (Chapter 3). Most studies that use Landsat data focus on regions of at most several hundred square kilometers. An example of a Landsat based study over a much larger area is the Large Area Crop Inventory Experiment (LACIE). LACIE was initiated to estimate crop yields in the former Soviet Union and to assess global grain markets more reliably. During LACIE, large amounts of Landsat data were used in combination with statistical sampling techniques and simple crop models to estimate crop production in the US and the former Soviet Union.

Continuous monitoring of vegetation at global scales became possible with the launch of the NOAA satellites in 1981. The NOAA satellites carry the Advanced Very High Resolution Radiometer (AVHRR), which collects data globally at least twice daily, once during daytime and once during the night (Chapter 3). However cloud cover and other atmospheric effects reduce the effective coverage. The NOAA satellite was designed for meteorological applications such as cloud analysis, analysis of sea surface temperatures and snow detection. NOAA-AVHRR data were first used for vegetation monitoring in the early eighties (a selection of early papers is referenced by Tucker 1996). The immediate motivation for this research was an imminent failure of Landsat 3 and its long overdue replacement by Landsat 4. AVHRR channel 1 and 2 data with 1.1 km resolution were combined in the Normalized Difference Vegetation Index (NDVI). Seasonal sums of NDVI were calculated from cloud free data over the growing season. The relationships between seasonal sums of NDVI versus annually accumulated above ground biomass (or crop yield) were similar to relationships established via ground measurements by Tucker (1979). Tucker and co-workers concluded that the AVHRR was suitable for the estimation of vegetation primary production. However, several shortcomings were also noticed: view angle dependent variations appeared in NDVI data, the spectral bandwidth of channels 1 and 2 was too broad, and the time of overpass was thought unsuitable: the 7:30 overpass of NOAA 6 was too early because of low light levels and the 14:30 time of overpass of NOAA 7 was thought too late because of cloud build-up.

The Nile-delta study was repeated for Senegal with NOAA 7 data (Tucker *et al.* 1983, 1985b) and for other parts of the Sahel with similar results (Justice 1986, Prince and Justice 1991). During the first year of the Senegal study Tucker and co-workers were fortunate that most of the AVHRR data were cloud free (Tucker *et al.* 1983). During the second year, cloud contamination, atmospheric effects, and variations in NDVI with viewing angle limited the usefulness of the data. Several techniques were investigated to reduce these effects and enhance data quality (Kimes *et al.* 1984; Holben 1986). The most suitable technique was to form 10 day, 15 day or monthly NDVI composites by selecting the maximum NDVI for each picture element (pixel) over the period considered. This maximum value compositing technique was implemented by the global inventory, modelling, and monitoring system (GIMMS) group to process AVHRR data into 10-day NDVI composites of Africa in near real-time. The United Nations Food and Agricultural Organization (FAO) started to use these data for famine early warning and desert locust detection (Hielkema 1990).

In 1983, NOAA started processing of the Global Vegetation Index (GVI) data set (Tarpley *et al.* 1984). The availability of the GVI and GIMMS-NDVI data sets prompted a host of new research on the spatial distribution and seasonal dynamics of vegetation over large areas. For example, Tucker *et al.* (1985a) estimated land-cover classes of Africa based on seasonal variations detected in NOAA-AVHRR data. Goward and Dye (1987) studied net primary production in North America. Malingreau *et al.* (1985) detected forest fires in southeast Asia during the 1982 El Niño drought. Helldén *et al.* (1984) studied land degradation in southern Sudan. Tucker *et al.* (1986b) studied the relationship between the annual cycles of NDVI and the seasonal variations in absorption of atmospheric CO_2 by vegetation.

When multi-year NOAA-AVHRR data sets became available, several shortcomings were noticed. These shortcomings are related to both the design of the AVHRR and to external sources. Design related shortcomings include changes in the sensitivity of the visible and near-infrared sensors, inaccuracies in the satellite navigation and a gradual change in the local crossing time of the satellite to later hours during its lifetime. External sources of interference are scattering and absorption of solar radiation by atmospheric constituents, variations in soil reflectance, clouds, and variations in NDVI with illumination angle and viewing angle. Illumination effects, also referred to as satellite drift effects, are noticeable as discontinuities in AVHRR time-series when the data receiving protocol changes from one NOAA platform to the next. Several techniques were developed to account for these inconsistencies. A change in sensitivity of the sensor, or sensor degradation, was corrected by assuming signals from deserts and cloud tops invariant (Rao and Chen 1994; Vermote and Kaufman 1995; Los 1998b). Orbital models were improved to obtain more accurate navigation of the NOAA satellites (Rosborough *et al.* 1994). Corrections were developed for atmospheric effects (Tanré *et al.* 1992; Vermote *et al.* 1995), cloud contamination (Stowe *et al.* 1991; Gutman *et al.* 1994), dependence of surface reflectance on viewing and solar angles (Roujean *et al.* 1992; Sellers *et al.* 1996b) and reflection from soil background (Qi *et al.* 1994). The implementation of some of these corrections is straightforward; others require data from additional sources that cannot be obtained with sufficient accuracy.

Several attempts were made to release corrected, spatially and temporally compatible data to the public. Amongst these data sets are the AVHRR land surface Pathfinder data set (James and Kalluri 1994), the FASIR NDVI data set (Sellers *et al.* 1996b; Los *et al.* 2000) and the 3rd generation GVI data set (Gutman *et al.* 1994). These data sets were developed for different purposes: The Pathfinder data and GVI data were developed for observational purposes whereas the FASIR NDVI data were designed for use with models.

With the availability of longer NOAA-AVHRR data records, it became possible to study the effects of interannual variations in climate on vegetation. An important source of climate variation is El Niño. El Niño is a warming of the eastern equatorial Pacific Ocean and has been associated with disruptions in rainfall patterns over large areas, especially in the tropics (Skidmore 1988). These disruptions in rainfall result in variations in vegetation and these can be detected from satellite (Myneni *et al.* 1995; Anyamba and Eastman 1996; Los *et al.* 2001). Warming of mid and high northern latitudes leads to earlier springs and this may result in an earlier greening of vegetation, although there is still debate about the accuracy of the reported trend in AVHRR NDVI data (Malmström *et al.* 1997; Myneni *et al.* 1997; Gutman 1999; and Los *et al.* 2000, 2001). A link between El Niño and wild fire activity has also been proven (Skidmore 1988).

Interest in the effects of vegetation on the global carbon balance, water balance and energy balance lead to the development of improved land surface sub-models in climate models and biogeochemical models (Sellers *et al.* 1996b, 1996c, Potter *et al.* 1993; Field *et al.* 1995). Satellite data are used to estimate biophysical parameters in these new land surface models (Sellers *et al.* 1996b, 1996c). Satellite derived biophysical land surface parameters show more realistic spatial and temporal variability than previously prescribed land surface parameters that were estimated from land cover types and look-up tables (Sellers *et al.* 1996b).

Useful information about vegetation cannot only be obtained from visible and near infrared channels, but also from thermal channels. Thermal channels have the ability to detect fires, if these are sufficiently large. Thermal data can thus provide information on timing and location of fires and provide information on the role of wild fires for, e.g. air quality and the carbon cycle. Interest in the study of wild fires renewed when large outbreaks occurred during the El Niño year of 1998.

5.3   LANDSAT BASED REGIONAL STUDIES

Landsat data are used mostly to obtain information of the land surface for a region. They provide a snapshot of a region for a particular time that is used to obtain information on land cover types (see also Chapters 3, 6, 7 and 9 for illustrations of several applications). For some areas two or three useful (e.g. cloud free) snapshots may be available for a particular year, other areas may have less frequent coverage. This low temporal coverage in combination with the large data volumes that need to be handled makes Landsat data less suitable for monitoring of large areas at frequent intervals. Because of the computing resources needed, uncertainties in acquiring data in a timely fashion and the costs involved, few institutions will be able to do high temporal resolution, large area analysis. Two examples follow where Landsat data are used to monitor vegetation over large areas. The first example uses statistical sampling of Landsat data to estimate crop yield in the former Soviet Union. The second example uses Landsat data at about 10-year intervals to estimate the rate of tropical deforestation in the Amazon.

5.3.1 The Large Area Crop Inventory Program

The Large Area Crop Inventory Program (LACIE) was initiated to obtain timely, reliable estimates of crop production for several important wheat exporting countries in the world (MacDonald and Hall 1980). These estimates are often lacking or inadequate because of insufficient funding or limited technical capabilities at the country level. For example, lack of information on a shortfall in grain crop in the former Soviet Union led the US to sell grain at a below-market price in 1972 and 1977. Timely availability of this information could have avoided this negative effect on the US economy.

The LACIE program used Landsat data, meteorological data (monthly mean temperature and rainfall) and ancillary data such as crop maturity calendars, cropping practices, field size, modelled adjustments to the normal wheat crop calendar in response to the current year's weather as well as summaries of the meteorological and crop conditions for the current crop year. Ground-acquired data on crop identification and crop condition were used to develop techniques and assess the accuracy of the LACIE system.

Crop acreage was estimated from stratified random samples of Landsat data with a manually assisted machine processing approach. This sampling was an important step because it reduced the large volumes of Landsat data to be handled. Stratification was based on administrative regions. Within each stratum, sample units were selected consisting of segments of Landsat data of approximately 9 by 11 km. About 2 per cent of the Landsat data were selected at random, which incurs a sampling error of less than 2 per cent. An image analyst used the spatial context in the segments and the phenological stage of vegetation to label about 100 Landsat pixels as small-grain crops or other crops. About 40 of the 100 labeled pixels were used to train clustering and maximum likelihood classification algorithms. These algorithms were used to classify the pixels of each of the other 23,000 segments. The remaining 50 labelled pixels were used to adjust the percentage of pixels computer-classified as small grain to estimate the proportion of the segment area where small grains were growing. Finally, the analyst verified the results of the classification.

Yield was estimated with multiple linear regression models of historical yields and monthly averages of temperature and precipitation:

Yield = A (yield for average weather) + B (adjustment for technology trend) + C (effects of current weather)   (1)

Yield models were dependent on region. Yield was estimated for larger regions and agricultural zones (e.g. the US Great Plains) by multiplying crop acreage with crop yield.

The LACIE project obtained good results for large homogenous agricultural fields such as those in the former Soviet Union and in larger regions in the US. The results were not as good for smaller fields, because the resolution of the Landsat data was too low. One should be able to improve upon the results for smaller areas by using multispectral sensors with higher spatial resolution (Chapter 3). LACIE demonstrated that remote sensing technology could be used to obtain reliable estimates of crop yield over large regions and identified some key problems in prediction of crop yield based on remote sensing technology. Consistency of estimates between years was obtained by training the satellite data with ground data.

5.3.2 Tropical deforestation and habitat fragmentation

Deforestation has occurred in temperate and tropical regions throughout history (e.g. Tucker and Richards 1983). In recent years attention has focused on tropical forests, where as much as 50 per cent of its original extent may have been lost to deforestation in the last two decades, primarily due to agricultural expansion (Myers 1991). Global estimates of tropical deforestation vary from 69,000 km^2 yr^{-1} in 1980 (FAO/UNEP 1981a, 1981b, 1981c) to 100,000 to 165,000 km^2 annually in the late 1980s. Of the more recent estimates, 50 per cent to 70 per cent have been attributed to deforestation in the Brazilian Amazon, the largest continuous tropical forest region in the world (Myers 1991).

Tropical deforestation is a component of the carbon cycle and has profound implications for biological diversity. Deforestation increases atmospheric CO_2 and other trace gases, possibly affecting climate (IPCC 1995). Conversion of forests to cropland and pasture results in a net flux of carbon to the atmosphere, because carbon contained in forests is higher than that in the agricultural areas that replace them. The lack of data on tropical deforestation limits our understanding of the carbon cycle and climate change (IPCC 1995). Furthermore, while occupying less than 7 per cent of the terrestrial surface, tropical forests are the home to half or more of all plant and animal species (Wilson *et al.* 1988). The primary adverse effect of tropical deforestation is massive extinction of species.

Deforestation affects biological diversity in three ways: destruction of habitat, isolation of fragments of formerly contiguous habitat, and edge effects within a boundary zone between forest and deforested areas (see also Chapter 7). This boundary zone extends some distance into the remaining forest. In this zone there is greater exposure to winds; dramatic micrometeorological differences occur over short distances; easier access is possible by livestock, other non-forest animals, and hunters; and a range of other biological and physical effects occurs. The cumulative results are loss of plant and animal species in the affected edge areas (Harris 1984).

There is a wide range in current estimates of the area deforested and the rate of deforestation in Amazonia. Brazilian scientists at the Instituto Nacional de Pesquisas Espaciais (INPE) have estimated Amazonian deforestation. These unpublished studies report a total deforested area of 280,000 km^2 as of 1988, and that the average annual rate from 1978 to 1988 was 21,000 km^2 yr^{-1}. More recently, INPE (1992) has reported deforestation rates for Brazil to range from 12,000 to 25,000 km^2 yr^{-1}. Other studies have reported rates ranging from 50,000 km^2 yr^{-1} to 80,000 km^2 yr^{-1} (Skole and Tucker 1993). Additional deforestation estimates have been made for geographically limited study areas in the southern Amazon Basin of Brazil using Landsat or meteorological data. This wide range of estimates clearly indicates a need for the timely availability of reliable, spatially contiguous data sets to monitor the rate of tropical deforestation. The Landsat data record that is available since the 1970s helps to fulfill this requirement.

5.4   AVHRR BASED REGIONAL AND GLOBAL STUDIES

AVHRR data are used for very diverse purposes. This section provides an overview of sources of interference in AVHRR data and a discussion on ways to detect and correct these. Several examples of applications are presented that are based on the ability to detect interannual variation in vegetation with AVHRR NDVI data, including one example that discusses the detection of wild fires with AVHRR thermal channels. If one would like to infer vegetation seasonality from NDVI data, very few corrections are needed since the seasonality is large (in temperate forests, typically ranging from 0.1 to 0.7). On the other hand, trends in NDVI data are small, in general much less than 0.05 NDVI over a period of 10 years, and are of the same order of magnitude as errors in carefully corrected data (Malmström *et al.* 1997). Between these two extremes is the detection of interannual variation with periods of 2 to 4 years. The change in NDVI is about 0.1; hence interannual variation can be determined from carefully corrected data.

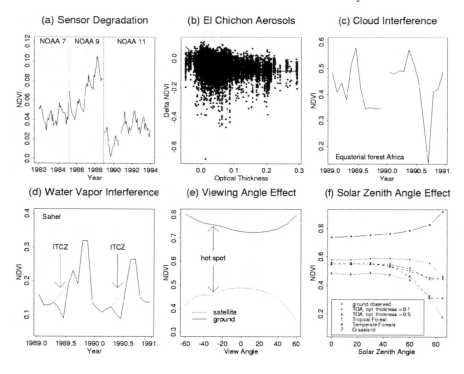

Figure 5.1. Interferences in NDVI data.

5.4.1 Sources of interference

A summary of interferences in AVHRR NDVI data is provided in Figure 5.1. Interferences are related to the sensor, the platform orbit, the atmosphere, and the land surface properties. These corrections are described in detail in Los *et al.*

(2000). Briefly Figure 5.1 shows (a) Variations in NDVI over the Saharan desert that result from calibration differences and variations in sensitivity of AVHRRs visible and near-infrared channels aboard NOAA-7, -9 and –11. (b) Decrease in NDVI with optical thickness; increased optical thickness is caused by aerosols from the El Chichon eruption. (c) Clouds result in lower or missing values over a tropical forest. (d) Decreased NDVI as a result of increased atmospheric water vapour at the beginning of the growing season in the Sahel. (e) Increase in NDVI with viewing angle for ground measurements (top line) and decrease in NDVI with viewing angle for top-of-the-atmosphere measurements. (f) Increase in NDVI with solar zenith angle for ground measurements (top line); decrease in NDVI with solar zenith angle for top-of-the -atmosphere measurements; + and x indicates different aerosol loading, 1 indicates relationship estimated from statistical analysis (from Los *et al.* 2000).

5.4.1.1 Sensor degradation

The sensitivity of the sensor changes over the time of operation of the NOAA satellite. As a result we see variations in NDVI time-series from areas such as deserts that we would otherwise expect to be invariant. Several approaches exist to correct for sensor degradation. All of these assume specific invariant targets and attribute changes in NDVI from these targets to sensor degradation; amongst these invariant targets are deserts (Rao and Chen 1994; Los 1998b), cloud tops (Vermote and Kaufman 1995), or maximum and minimum statistics collected for the entire globe (Brest *et al.* 1997). Calibration is performed in two steps; first the so-called pre-flight calibration is applied; this is the calibration that is done prior to launch. A calibration offset can also be estimated from measurements that the AVHRR takes from outer space. This so-called deep space count provides a more reliable indication of the offset than the preflight calibration offset.

A relative calibration can be obtained by selecting an arbitrary data point of a time series as a reference; absolute calibration can be obtained by using additional information, e.g., from aircraft, to adjust any point of the time series, and then adjust other points accordingly.

A slight complication is that in general deserts do exhibit some seasonality. This seasonality is most likely caused by variations in water vapour content; channel 2 is sensitive to variation in water vapour. By subtracting the seasonal cycle from the estimated r1/r2, this effect can be significantly reduced. Solar zenith angles do not significantly affect the top-of-the-atmosphere NDVI over the Sahara desert (Los 1998b).

5.4.1.2 Atmospheric effects and correction (Figure 5.1.b and d)

The atmosphere alters the radiation reflected from the land surface by absorption and scattering. Atmospheric water vapour and ozone are strong absorbers of solar radiation, whereas molecules and aerosols predominantly scatter (Chapter 3).

Scattering by molecules and absorption by ozone can be accounted for by using topographic data (molecular scattering) or an ozone climatology or observations from satellite (ozone). The effects of aerosols and water vapour are more difficult to account for, since their concentrations are highly variable in space and time. Aerosols increase scattering by the atmosphere in both the visible and near infrared bands. The effect of aerosols is strongest over dark dense vegetation; for these conditions the visible reflectance increases relative to the near infrared reflectance and this causes a decrease in the NDVI.

Aerosols can originate from deserts or bare soils, these aerosols remain for the most part in the troposphere, or they can originate from volcanic eruptions; part of the volcanic aerosols can reach the stratosphere. Data on tropospheric aerosols are hard to obtain and it is therefore difficult to correct for these.

Stratospheric aerosols from volcanic eruptions can remain airborne for several years. The effects of the eruptions by El Chichon (Mexico) in April 1982 and of Mt Pinatubo (the Philippines) in 1991 lingered for several years. Initially the distribution of aerosols is highly concentrated. Several months after an eruption, stratospheric aerosols are well mixed, and their concentration depends mostly on latitude and time since the eruption. The effect of aerosols can be estimated from observations over dark targets such as open water (oceans) or dark dense vegetation. Figure 5.2.b shows a plot of optical thickness values that were measured after the eruption of El Chichon in 1982 in Mexico from AVHRR data over the Pacific Ocean (Vermote *et al.* 1997) versus deviations in NDVI from areas with a mean monthly NDVI > 0.5. This relationship was used to correct the FASIR NDVI data for the effects of volcanic aerosols (Los *et al.* 2000). The aerosols affect the NDVI in a near-linear way, with the largest effects on high NDVI values from dense vegetation and negligible effects on low NDVI values from bare soils. A simple approximation of this relationship is

$$\Delta NDVI = \Delta NDVI_{max}(NDVI-NDVI_{min})/(NDVI_{max}-\Delta NDVI_{max}-NDVI_{min}) \qquad (2)$$

where $\Delta NDVI$ is the expected change in NDVI as a result of volcanic aerosols; $\Delta NDVI_{max}$ is the maximum change in NDVI as a function of latitude and time; NDVI is the measured contaminated NDVI; $\Delta NDVI_{max}$ is the maximum NDVI for a particular latitude; and $NDVI_{min}$ is the minimum NDVI (minimum NDVI value over deserts). Estimates of these values are data set dependent.

5.4.1.3 Clouds

Clouds obscure the land surface and are especially a problem over areas with frequent rainfall and associated dense vegetation (Figure 5.1.c). Clouds have low NDVI values. The maximum value compositing, i.e. selecting the maximum NDVI over a period of 10 days, 15 days or a month, reduces the effects of clouds by selecting data from clear, cloud-free days with high NDVI values. In extreme cases, e.g. over tropical forests, very few cloud-free data may be available. In these cases the compositing period could be expanded (e.g. to two months), or spatial filters could be used that select the maximum NDVI over a moving window of 3x3 or 5x5 pixels. The spatial or temporal resolution decreases, but a more representative value is likely selected.

Short-term variations in NDVI data that result from atmospheric aerosols or clouds can be corrected by using information from the entire time-series. One way to reduce the effect of these outliers is by fitting Fourier series through the data with linear regression. By using a weighted least squares solution, spurious data points are identified and replaced with estimates from the fitted Fourier series (Sellers *et al.* 1996b).

5.4.1.4 Solar zenith angle effects

The NDVI varies with the solar zenith angle, i.e., the angle that the sun makes with a line normal to the Earth's surface. The variation depends on the type of land surface and the atmosphere. The NDVI of dark dense vegetation tends to increase with increasing solar zenith angle when measured at the ground. At the top of the atmosphere, larger solar zenith angles lead to increased scattering and the NDVI of dark dense vegetation tend to decrease with solar zenith angle (Figure 5.1.f). Hence the effects of atmosphere and land surface in part compensate each other. The effect of the atmosphere is in general stronger, hence NDVI of dark dense vegetation measured at the top of the atmosphere tends to decrease with solar zenith angle. Solar zenith angle effects in the NDVI of bare soils tend to be small, if not negligible. The direction of solar zenith angle effect of NDVIs from sparse vegetation measured at the top of the atmosphere varies dependent on land surface properties and composition of the atmosphere. Sellers *et al.* (1996b) assume the effect to be linear with NDVI, which seems valid for some sparsely vegetated areas, but not for others.

The solar zenith angle depends on latitude, time of year and time of day. The effect of solar zenith angle on NDVI data can be estimated from a statistical analysis of NDVI distributions. First the land surface is stratified into classes with equal solar zenith angle interval and vegetation type – only vegetation types that can develop fully green conditions should be selected. NDVI histograms are calculated for each of these classes. The maxima of these histograms represent fully green conditions. The solar zenith angle effect can than be established by plotting per vegetation class, the 98 or 95 per cent of the NDVI versus the solar zenith angle interval. The green conditions represented by the 98 per cent values show a decrease in the NDVI with increased solar zenith angle (Figure 5.1). An equation for top-of-the-atmosphere solar zenith angle effect valid between solar zenith angle of 30° to 60° is given by

$$NDVI_{98} = NDVI_{98,0} - k_1(\theta - \pi/6)^{\wedge}k_2 \qquad (3)$$

Where $NDVI_{98,0}$ is the NDVI for an overhead sun, θ is the solar zenith angle, and k_1 and k_2 are constants that can be determined with non-linear regression. Equation 3 can also be log transformed and the constants can than be estimated with linear regression. For low vegetation conditions the analysis can be repeated for NDVI values from bare soils using the 2 percentiles instead of the 98 percentiles. Effects of solar zenith angle on the NDVI of bare soils are in general small. After establishing the effects for minimum and maximum NDVI conditions, the data can be corrected assuming a linear effect of solar zenith angle on intermediate NDVI values

$$NDVI_0 = (NDVI_\theta - NDVI_2)(NDVI_{98,0} - NDVI_2)/(NDVI_{98,\theta} - NDVI_2) + NDVI_2 \qquad (4)$$

Gutman (1999) also proposed a correction of NDVI data based on variations in low and high NDVI values, but derived land cover classes from the seasonality of NDVI time-series.

5.4.1.5   Soil background variations

Variations in reflective properties of the bare soil can result in variations in NDVI that are not related to variations in vegetation. In addition, evidence has been found that variations in atmospheric composition, e.g. water vapur, result in temporal variations in NDVI from bare soils (Rao and Chen 1994). These effects are of importance for studies in desert margins. We will find two examples that deal with soil background effects on NDVI (sections 5.4.2 and 5.4.3).

5.4.1.6   Discussion

The corrections for sensor degradation, solar zenith angle effects and volcanic aerosols are the most important ones for the study of interannual variation and are essential in many cases to do meaningful comparisons between years. Other corrections may be applied as well; corrections for ozone absorption and Rayleigh scattering will bring NDVI values closer to ground measurements. However, these corrections may come at a cost when daily data are composited; maximum value compositing on atmospherically corrected data tends to select data from off nadir-viewing angles because these are higher; maximum value compositing on uncorrected data tends to select data from near-nadir viewing angles (compare top and bottom lines in Figure 5.2.e). To avoid problems that are associated with selection of data from off-nadir viewing angles (such as a larger pixel size, overestimation of vegetation greenness, and large differences in observed vegetation greenness for slight variations in viewing angle) a BRDF correction is needed prior to maximum value compositing. The range of solar zenith angles and viewing angles from which AVHRR data are selected over a compositing period is in many cases insufficient to do a proper BRDF correction.

   In the next couple of sections several AVHRR NDVI based applications are discussed.

5.4.2 Desert margin studies

The term desertification is most often used to describe the advance of desert-like conditions (UNCOD 1977). Poor land management coupled with increased population are often blamed for catastrophic desertification in semi-arid regions such as sub-Saharan Africa. In the mid-seventies several authors reported an expansion of the Saharan desert estimated at a rate of 5.5 km per year (Lamprey 1975). This desert encroachment had direct consequences for humans, animals and plants, and could trigger a positive feedback loop on the (local) climate that would further increase the desert expansion (Charney *et al.* 1977). This positive was by an initial decrease in vegetation. Because vegetation is in general darker than the underlying soil, the total albedo of the surface and less solar radiation is absorbed by the land surface. The total energy emitted by the land surface in the form of latent and sensible heat to the atmosphere is reduced, and this reduces evapotranspiration and convection. Rainfall is reduced as a result of lower concentrations of moisture in the atmosphere and less moisture is available for plant growth. The initial reduction in vegetation thus causes a further decrease in the vegetation amount, or at least sustains itself. A similar argument can be made for the reversed mode of the feedback loop, i.e. more vegetation leads to lower albedo, more heat absorption by the land surface, higher evaporation, higher rainfall amounts and more vegetation. A positive feedback loop is by its very nature highly unstable and it is therefore obvious that the positive feedback loop between climate and vegetation is not the only mechanism at work in the Sahel. Large weather systems such as the seasonal movements of the Inter Tropical Convergence Zone (ITCZ) that are driven by warming and cooling of the oceans dominate the Sahelian rainfall. Nevertheless, the mechanism is likely to enhance variations in the local climate near desert margins however, and for this reason, expansions and contraction of desert margins are thought to be an important indicator of global climate change.

Helldén and co-workers (Helldén 1984, 1988; Olsson 1985) questioned the rate of desertification of 5.5 km per year estimated by Lamprey (1975). They showed that the high rates of desertification were derived from local areas and were not valid for the entire Sahel. Moreover, the climate in the Sahel is characterized by cycles of increasing and decreasing wet and dry conditions (Nicholson *et al.* 1988). The study of Lamprey (1975) compared data from a wet and a drought cycle and thus overestimated the rate of desert encroachment. Tucker *et al.* (1991) studied the interannual variation of AVHRR based NDVI data to investigate the rate of desertification for the entire Sahel.

The study of Tucker *et al.* (1991) used monthly GIMMS NDVI composites for Africa. The data were corrected for sensor degradation and soil background effects to determine more accurately the interannual variations in NDVI. The NDVI was corrected for sensor degradation correction by adding monthly offsets, ΔNDVI

$$NDVI_d \quad = \quad NDVI_p + \Delta NDVI, \qquad (5)$$

where $NDVI_d$ and $NDVI_p$ stand for preflight and desert calibrated NDVI respectively. Equation 5 is an adequate approximation for sensor degradation between NDVI values of 0 and 0.3 (Los 1993). Solar zenith angle effects and volcanic aerosol effects are likely to be small because NDVI values are low. The soil background effect was estimated from areas with a standard deviation over the growing season smaller than 0.04 NDVI. For these areas NDVI values corresponding to the highest channel 5 brightness temperatures during the end of the dry period (1 to 21 March) of 1989 were subtracted from the annual time-series. NDVI data were then summed over the growing season and a positive relationship was derived between growing season mean NDVI and annual rainfall using linear regression. By defining the Sahelian zone as the region with an average annual precipitation between 200 and 400 mm and relating this number to a seasonally summed NDVI, it became possible to detect year-on-year changes in the aerial extent of vegetation in the Sahel. These variations showed an expansion of desert like conditions in the Sahel from 1981 to 1984, which was a year of severe drought. After 1984, conditions improved which resulted in a gradual increase in the amount of vegetation in the Sahel. Superposed on this long-term trend are interannual variations in vegetation extent caused by a interannual variations in rainfall. Ellis and Swift (1988) reported that these areas are in non-equilibrium by nature, i.e. they are perturbed by abiotic forces, usually droughts.

The results from the study by Tucker clearly demonstrate that in order to assess desertification in the Sahel, data must be interpreted in relation to long-term climatic cycles of drought and wetness. It is then possible to estimate the effects of localized degradation and erosion as deviations from a long-term trend. Several authors (Helldén 1988; Olsson 1985; Prince 1991) concluded that increased rangeland activity and intensified agricultural practices led to, e.g. a shift in vegetation composition towards species not favoured by cattle, shorter periods between fallow and cultivated cycles, increased erosion, and decreased infiltration capacity of the soil. They did not find evidence for a rapid expansion of the Saharan desert.

5.4.3 Monitoring Desert Locust habitats

The Desert Locust, *Schistocerca gregaria*, is still a common threat to the vulnerable pastures, subsistence farming and agriculture in the arid and semi-arid areas of Northern and Sahelian Africa, the Middle East and southwest Asia. The control of the Desert Locust is an important part of the general effort in ensuring food security in these regions. The Food and Agriculture Organization of the United Nations (FAO) adopted a preventive control approach aimed at reducing cost, scale and environmental hazards. Early warning is the cornerstone of this preventive strategy. Locust situation evaluation and forecasting at national and global scale is a prime activity of the early warning. To be successful, timely and reliable information on various locust population and environmental parameters, such as rain and vegetation, is required.

Traditionally, Plant Protection Services organize field surveys to estimate the condition of the locust habitats, in terms of soil moisture and vegetation, and the situation of locust populations. However, field surveys are expensive, cover only small areas and cannot be frequent. Low resolution satellite remote sensing offers a more cost-effective means to obtain a synoptic overview on the ecological conditions of the natural habitat of the Desert Locust. This information helps to forewarn of suitable breeding conditions, target field surveys and thus improves their efficiency.

Rainfall is a first indicator of a potential favourable condition of the locust habitat. But there are hardly any representative weather stations and estimates from satellites are still not adequate. Vegetation development that follows rain is required by locusts for food and shelter and can be observed by satellite. In terms of cost, time frequency and area coverage, NOAA AVHRR data are very suitable for this routine monitoring task. Real time data are required. Historic data are of limited value as the rain events, triggering the vegetation growth, are too erratic in time and space in this desert environment to make historic comparison sensible. Furthermore, vegetated locust-breeding areas are rather small, mostly not more than a few square kilometers. When applying NOAA GAC (4 km) data, it was experienced that many of these vegetated patches, important though for locust breeding, were not detected at this coarse resolution. The 1 km pixel size of the NOAA LAC/HRPT data proved to be more adequate.

Projects at FAO want to make the NOAA AVHRR vegetation index (NDVI) (Tucker *et al.* 1984 and 1986b) products more applicable to the non-technical locust control staff. This includes the development of methods to increase the reliability of the NDVI as well as providing the tools for integrating the NOAA NDVI with other spatial information, such as maps on soil conditions, potential vegetation or locust suitability, to improve the final interpretation.

In relation to monitoring locust-breeding areas from an early warning perspective, Hielkema *et al.* (1986) proposed an analytic approach by synthesizing, based on a geographical grid, single NDVI values, above a certain threshold, into a factor that would represent potential suitable conditions for breeding. This approach took into consideration the restricted computing power on the user's side at that time. Application of the method quickly shed light on one of the basic shortcomings of the NOAA NDVI. The unpredictable variation in space and time of NDVI values imposing the inability of defining reliable thresholds to distinguish soil from vegetation consistently over one image and throughout a season.

The reliability of the NDVI applied for locust habitat monitoring is hampered by a weak reflectance of the low plant cover, typical for the desert breeding areas. This enhances the problem of soil reflectance interference on the NDVI. Variations in soil reflectance and spectral characteristics of quartzite desert soils, affect the vegetation index. Improvement of the NOAA NDVI, to represent truly the vegetation cover on the ground, can be obtained by standardizing the background reflection (Cherlet and Di Gregorio 1993). An insight this work provoked was the consideration that by applying similar procedures the NDVI can be somehow improved, strongly depending on the geo-physical aspects of the area, but that limits are now reached in terms of NOAA sensor and data quality and sensitivity for desert vegetation monitoring. Improvements can be expected when satellite sensors with spectral bands more adapted to vegetation detection and better geometric correction of images become available at the same 1 km resolution. However, high spatial resolution imagery indicated that further improvements would be reached with higher resolutions than 1 km, such as 500 m or even 250 m. Similar time frequency, area coverage and cost of data would be required, though, to keep the same advantages with regard to cost-efficient monitoring.

Apart from these tentative improvements to the NOAA NDVI data itself, the routine and operational interpretation of the NDVI for desert habitat monitoring is made more reliable by integrating the data in a GIS application. An integrated information system, RAMSES (Reconnaissance And Management System for the Environment of *Schistocerca* i.e. Desert Locust), was developed to support the national locust early warning efforts. A custom-made user interface makes it easy for the non-specialists to access a variety of data, such as locust information, rainfall, locust habitat data and remote sensing products, such as the NDVI. The land unit map contains a linked database inventorying the important environmental parameters. This spatial information is an aid to a more correct interpretation of the NOAA NDVI in view of evaluation of locust habitat conditions.

The main impact of using routinely NOAA AVHRR NDVI products integrated with field data on locust and their habitat and meteorological data is a significant contribution to decision-making procedures for Desert Locust survey and control planning. Timing of surveys can be optimally planned. Surveys can be reduced to cover habitats estimated to be suitable for breeding. Early detected potential hazard areas can be monitored, increasing the preparedness for intervention. Subsequent control operations become sharply targeted. This can save time, available resources and reduces hazardous environmental impact of chemical campaign control efforts.

5.4.4 Land cover classification

The use of satellite data for the classification of land cover types is a useful addition to classical methods that use observations from ground surveys. In general, satellite based land cover classifications have used Landsat and SPOT imagery. Most authors claim classification accuracies of between 60 per cent and 90 per cent, dependent on the scale of the object being mapped (for example, using the Anderson scale, are you mapping forest from soil, coniferous forest from deciduous forest, or specific forest types such as oak or aspen), availability of training data and separability of classes (Skidmore *et al.* 1988). Land surface classifications from a combination of satellite and ground data can be obtained at a lower cost when compared to classifications solely based on ground surveys and aerial photograph interpretation. Landsat and SPOT based land cover classifications generally cover areas the size of one or a few scenes, the size of one Landsat scene being about 185 by 185 km.

Land cover classifications for areas the size of continents or for the entire globe are used for different purposes than the classifications from Landsat or SPOT scenes and generally require less detail both in the distinction of the number of classes and the spatial resolution of the data. These classifications, based on NOAA AVHRR images, are used to provide input parameters to mesoscale and global scale climate and carbon models (Dorman and Sellers 1989; Henderson Sellers and Pitman 1992; Field *et al.* 1995). Examples are the classifications by Matthews (1983), Kuchler (1983), and Wilson and Henderson-Sellers (1985) derived for climate models and by Olson *et al.* (1983) derived for carbon models (see also Chapter 6). These global classifications are compiled from ground based sources such as atlases, local maps, and reports. The output from these classifications often have large inconsistencies and errors because of differences in class definitions and interpretation, inclusion of outdated material, differences in spatial detail and scale, and the purpose for which the material was derived.

For example, Tucker *et al.* (1985a) made a land cover classification of Africa based on the first two principal components of three week composites of the NOAA-GVI data set over a 19 month period (April 1982 until November 1983). The first principal component was related to the average NDVI over the year, and the second principal component was related to the magnitude of the annual cycle. The results of the classification show a qualitative agreement with the major vegetation types found by White (1983), although deviations were found in several areas. Tucker *et al.* (1985a) suggested refinements of the method and additional years of data to improve the classification.

DeFries and Townshend (1994) provide a systematic discussion of limitations to land cover classifications based on NDVI time-series. They found that temporal profiles may not always be distinguishable, e.g., similar land cover types may have different temporal profiles and different land cover types may have similar profiles. Another complication is that identical land cover types may have similar temporal NDVI profiles but with a shift in the phase as a result of differences in the timing of the seasons. The reversed seasonality on the Northern and Southern Hemispheres is an example.

As a first approach to global classifications based on NDVI time-series, data are analyzed by region or by continent. This, of course, has limitations because it does not address climate gradients within a region and it limits the number of pixels for training areas and may give less reliable results. A second approach is to use additional information, for example data could be stratified by latitude and data could be classified by latitude interval. The strong latitudinal dependency of the occurrence of land cover types will lead to a small loss of training sites per class. The added information helps to distinguish between classes with otherwise similar profiles. A third approach is to standardize the timing of the temporal profiles, by shifting the profiles until the month of maximum NDVI is the same for all time-series. A fourth, in some ways related approach, is to run the classification on temporal characteristics of the NDVI time-series such as the average annual NDVI, the maximum range in NDVI values during the year, the length of growing season, and the rapidity of the rate of greening and senescence. These latter two techniques, without further information, do not solve the problems of distinguishing different vegetation classes with a similar temporal profile. To improve classifications, Loveland *et al.* (1991) used several measures of NDVI time-series (maximum NDVI, start of growing season, month with maximum NDVI and so on) in combination with climate data (mean precipitation and temperature), elevation, ecoregions and land resource data.

The identification of reliable training sites for global land cover classifications is an important issue. DeFries and Townshend (1994) used three data sets (Matthews 1983; Olson *et al.* 1983; Wilson and Henderson-Sellers 1985) and identified areas where there was agreement between these data sets. An alternative way to obtain training sites is from the interpretation of SPOT and MSS imagery, and to scale the results to the size of AVHRR pixels or global 1° by 1° grid cells. The reliability of the classification can be assessed by the spectral separability of the classes from training sites, and by testing the classification on sites not used for training the classes. Sensor degradation effects, solar zenith angle effects and volcanic aerosol are of minor importance because firstly, only one year of data is analyzed at the time and secondly, the AVHRR classification is directly trained with interpretations of Landsat data. These effects, as well as the uncertainties in the classification, do become serious issues if one would like to compare classifications from different years.

Results of NDVI based land cover classifications by Tucker *et al.* (1985a), Loveland *et al.* (1991) and DeFries and Townshend (1994) showed promise but also revealed that the accuracy is not as good as the accuracy of MSS or SPOT based classifications. However, Loveland *et al.* 1991 reported improvements in large area classifications with AVHRR data compared with large area classifications from conventional ground surveys. Further improvements should be obtained by incorporation of additional data, for example additional spectral channels, BRDF properties of the land surface, and climate data, and by use of higher spatial and spectral resolution data, such as from the MODIS instrument.

5.4.5 ENSO

Interannual variations in NDVI data are linked to interannual climate perturbations such as the El Niño Southern Oscillation (ENSO) phenomenon. The NDVI record of 18 years provides a unique basis for studying climate variability in areas where climatic data such as rainfall data are sparse.

One important aspect in understanding the response patterns of the land biosphere to interannual climate variability is in determining regions of the world that show a marked response to climate variability. The ENSO phenomenon is a principal cause of global interannual climate variability. ENSO events are triggered by the anomalous warming of oceanic waters in the central to eastern Pacific Ocean. The warming of the eastern Pacific is associated with lower than normal barometric pressure at Tahiti and higher than normal pressure over Darwin, Australia. This seesaw in sea level pressure between these two stations, termed Southern Oscillation Index (SOI) is the most commonly used index for ENSO phenomena and permits the identification of warm and cold ENSO events back to the late 19th century. Strong negative anomalies are associated with weakening easterly winds and a sea surface temperature warming of equatorial Pacific waters off the South American coast which is also referred to as an 'El Niño' event. The reverse of these conditions or cooling of oceanic waters in the eastern Pacific is termed 'La Nina'. Anomalous climatic conditions caused by ENSO such as increased rainfall or occurrence of drought is linked with outbreaks of human and livestock diseases in various countries. During the warm phase of ENSO there is a tendency for above normal rainfall in the eastern Pacific and below normal rainfall in the western Pacific Belt and eastern Indian Ocean. The climatic conditions are typically opposite during cold ENSO events or La Ninas. Over Africa there is tendency for differential hemispheric response patterns, with Eastern Africa receiving above normal rainfall and dry conditions prevailing over Southern Africa during warm ENSO events and the opposite effects during cold events. Occurrences of the warm ENSO phase have been linked to increased rainfall, increased vegetation growth and outbreaks of Rift Valley Fever in East Africa.

Principal component analysis is a powerful way to reduce data volumes and find associations between variables in the data. Principal components analysis in the time domain has been used to detect seasonal signals (Eastman and Fulk 1993;

Andres *et al.* 1994) and interannual signals in vegetation that are related to ENSO (Eastman and Fulk 1993; Anyamba and Eastman 1996).

Figure 5.2 shows a comparison between principal components loadings from a PCA analysis of a monthly NDVI time series for Africa for the period 1986 – 1990. The comparison with Sea Surface Temperature (SST) anomalies from the eastern Pacific NINO3.4 region shows an inverse relationship with NDVI variability in Southern Africa detected by the PCA. During the 1986/87 warm ENSO event, SSTs in the Pacific are warmer than normal while NDVI is below normal (negative loadings) because of below normal rainfall in southern Africa.

During the cold phase (La Nina), SSTs in the eastern Pacific are below normal and NDVI over Southern Africa is above normal. This pattern is illustrative of ENSO teleconnections, where by changes in the ocean-atmosphere coupling in the Pacific, impact rainfall and thus vegetation greenness at distant or remote locations from the Pacific (Ropelewski and Halpert 1987, 1989).

Rainfall data have incomplete spatial coverage and show gaps in temporal records. Satellite vegetation index data provide contiguous spatial coverage and complete time-series. This continuous coverage allows monitoring of the spatial and temporal development of a drought. This may provide important clues about the outbreak of tropical diseases that are related to extremes in climate (Nicholls 1991).

For example it has been shown that Rift Valley Fever (RVF) outbreaks in East Africa are associated with periods of negative SOI (El Niño) periods associated with heavy rainfall (Figure 5.3).

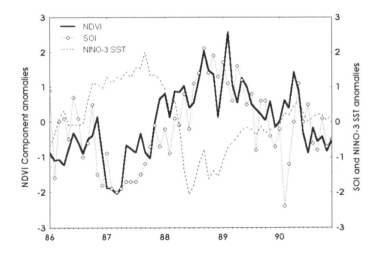

Figure 5.2: Comparison between NDVI component anomalies for Southern Africa against NINO3.4 SST and SOI anomalies for the period 1986–1990.

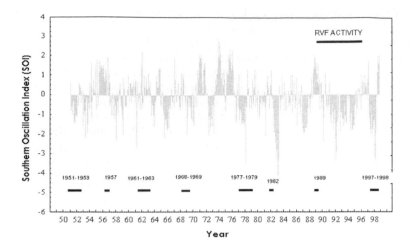

Figure 5.3: RVF outbreak events superimposed upon the Southen Oscillation Index (SOI) for the period 1950–1998. Outbreaks have a tendency to occur during the negative phase of SOI which is associated with above normal rainfall over East Africa.

Such heavy rainfall floods mosquito breeding habitats in East Africa, known as *dambos*, that are good habitats for *Culex* species mosquito vectors that transmit RVF. Areas that are experiencing anomalous climatic conditions, in this case increased rainfall and associated vegetation greenness, can be identified with NDVI time series (Linthicum *et al.* 1999).

5.5   WILD FIRE DETECTION

The detection of high-temperature sources using the AVHRR was first proposed by Dozier in 1981 (Dozier 1981; Matson and Dozier 1981). Since that time, the greater appreciation of biomass burning as a major factor in the emission of trace gases, and the resulting effect upon climate change, has led to an extensive body of research concerning satellite-based fire detection (Langaas 1995).

A typical land surface, which is approximately a 300 K black body, emits far more energy at 10.8 μm than at 3.8 μm. At typical smoldering and flaming temperatures of 500 to 1200 K, however, the emitted energy at 3.8 μm greatly exceeds that emitted at 10.8 μm. A subpixel high temperature source, such as a fire, will consequently elevate the observed response in NOAA AVHRR channel 3 compared to that of channel 4. This characteristic difference in response is exploited when looking for active fires.

In practice, AVHRR-based fire detection is complicated by the fact that other phenomena unrelated to burning produce the same characteristic signature. The most significant of these is reflected sunlight, which has a substantial 3.8 μm component. The relatively low saturation levels of the thermal bands (for land) further complicate AVHRR-based fire detection.

Most of the existing operational AVHRR fire detection algorithms have been developed for regional application and have been tuned accordingly. In light of the extensive range of biomes in which biomass burning may occur, however, more recent efforts have focussed on the development of detection algorithms that may be applied globally (Giglio *et al.* 1999). Generally, the techniques utilize similar processing steps and input data, and can be placed into two broad categories: fixed-threshold techniques and contextual techniques (Justice and Dowty 1994). Fixed threshold algorithms generally rely on preset absolute thresholds and consider a single pixel at a time, while contextual methods compute relative thresholds based upon statistics calculated from neighboring pixels.

As the majority of remotely sensed fire fronts are less than 1 km^2 in size, most AVHRR fire detection algorithms use 1 km LAC or HRPT data. In some instances GAC observations have been found to be useful (Koffi *et al.* 1996), but the severe sampling and averaging used to create the coarser resolution GAC data eliminates a significant number of fire pixels that would otherwise be detected.

5.6 DISCUSSION

Remote sensing of the global biosphere has developed from a qualitative science towards a quantitative science. During the early years satellite data were used as an observational tool to obtain information about the distribution of different types of land cover. AVHRR data allowed monitoring of the seasonality of vegetation. Comparison of AVHRR data between years created a need for corrections of the data. Newer satellite systems were designed to avoid many of the problems inherent with using AVHRR imagery. Improved calibration of data and improved corrections permit more straightforward comparisons between years.

The long-term record of Landsat and AVHRR data has provided us with unique insights in the functioning of the global biosphere and provided us a better understanding of seaonality of vegetation around the globe, rates of tropical deforestation and land cover change. We show several examples of studies that used either high spatial resolution or high temporal resolution data. In some cases a need exists for both; several applications indicated that the 4 km pixel size was a limitation (e.g. Locust breeding, fire detection). Better results are expected from higher resolution sensors with at least the same recurrence time as the AVHRR (e.g. Terra-MODIS, ENVISAT). Other applications, such as detection of an earlier growing season, do not require higher spatial resolution data, but would benefit from higher temporal resolution data. More frequent coverage would increase the probability of obtaining cloud-free data for a shorter period of time and would more accurately establish the start and end of the growing season.

5.7 REFERENCES

Andres, L., Salas, W. A., and Skole, D., 1994, Fourier analysis of multi-temporal AVHRR data applied to a land cover classification. *International Journal of Remote Sensing*, **15**, pp. 1115-1221.

Anyamba, A. and Eastman, J.R., 1996, Interannual variability of NDVI over Africa and its relation to El Nĩo Southern Oscillation, *International Journal of Remote Sensing*, **17**, 2533-2548.

Brest, C.L., Rossow, W.B. and Roitier, M.D., 1997, Update of radiance calibrations for ISCCP, J. Atmos. *Oceanic Technolo.* **14**, 1091-1109.

Charney J., Quirk, W.J. Chow, S.H. Kornfield, J., 1977, Comparative study of effects of albedo change on drought in semi-arid regions, *Journal of Atmos. Sciences*, **34**, 1366-1385.

Cherlet M.R., and Di Gregorio, A., 1993, *Calibration and Integrated Modelling of Remote Sensing Data for Desert Locust Habitat Monitoring*, FAO RSC series N. 64, 115pp.

DeFries, R.S., and Townshend, J., 1994, NDVI-derived land-cover classifications at a global scale, *International Journal of Remote Sensing*, **15**, 3567-3586.

Dorman, J. L., and Sellers, P. J., 1989, A global climatology of albedo, roughness length and stomatal resistance for atmospheric general circulation models as represented by the simple biosphere model (SiB). *Journal of Applied Meteorolgy*, **28**, 833-855.

Dozier, J., 1981, A Method for Satellite Identification of Surface Temperature Fields of Subpixel Resolution. *Remote Sensing of Environment*, **11**, 221-229.

Eastman R.J., and Fulk, M., 1993, Long sequence time-series evaluation using standardized principle components, *Photogrammetric Engineering and Remote Sensing*, **59**, 1307-1312.

Eischeid, J.K., Diaz, H.F. Bradley, R.S. and Jones, P.D., (eds), 1991, *A comprehensive precipitation dataset for global land areas*. U.S. Dept of Energy, Washington D.C., 82 pp.

Ellis, J.E. and Swift, D.M., 1988, Stability of African pastoral ecosystems: Alternate paradigms and implications for development. *Journal of Range Management*, **41**, 450-459.

FAO/UNEP, 1981a, *Los Recursos Forestales de la America Tropical*, 32/6.1301-78-04, Technical Report No. 1. Rome, Food and Agriculture Organization of the United Nations.

FAO/UNEP, 1981b, *Forest Resources of Tropical Africa*, 32/6.1301-78-04, Technical Report No. 2. Rome, Food and Agriculture Organization of the United Nations.

FAO/UNEP, 1981c, *Forest Resources of Tropical Asia*, 32/6.1301-78-04, Technical Report No. 3. Rome, Food and Agriculture Organization of the United Nations.

Field, C.B., Randerson, J.T. and Malmstrom, C.M., 1995, Global net primary production: combining ecology and remote sensing. *Remote Sensing of Environment*, **51**, 74-88.

Fung I.Y., Tucker, C.J. and Prentice, K.C., 1987, Application of AVHRR vegetation index to study atmosphere-biosphere exchange of CO_2. *Journal of Geophysics Resources*, **92**, 2999-3015.

Giglio, L., Kendall, J., and Justice, C. O., 1999, Evaluation of Global Fire Detection Algorithms Using Simulated AVHRR Infrared Data. *International Journal of Remote Sensing*, **20**, 1947-1985.

Goulden, M.L., Munger, J.W. Fan, S.M. Daube, B.C. and Wolfsy, S.C., 1996, Exchange of carbon dioxide by a deciduous forest: Response to interannual climate, *Science*, **271**, 1576-1578.

Goulden, M.L., Wofsy, S.C. Harden, J.W. Trumbore, S.E. Crill, P.M. Gower, S.T. Fries, T. Daube, B.C. Fan, S.M. Sutton, D.J. Bazzaz, A. and Munger, J.W., 1998, Sensitivity of boreal forest carbon balance to soil thaw, *Science*, **279**, 214-217.

Goward S.N., and Dye D.G., 1987, Evaluating North American net primary productivity with satellite observations, *Advanced Space Resources*, **7**, 165-174.

Gutman, G.G.. 1999, On the use of long-term global data of land reflectances and vegetation indices derived from the advanced very high resolution radiometer, *Journal of Geophysics Resources*, **104**, 6241-6255.

Gutman G.G., Ignatov, A. and Olson, S., 1994, Towards better quality of AVHRR composite images over land: reduction of cloud contamination. *Remote Sensing of Environment*, **50**, 134-148.

Hall, F.G., Huemmrich, K.F. Goetz, S.J. Sellers, P.J. and Nickeson, J.E., 1992, Satellite remote sensing of surface energy balance: Success, failures, and unresolved issues in FIFE. *Journal of Geophysics Resources*, **97**, 19061-19089.

Hall, F.G., Sellers, P.J. Apps, M. Baldocchi, D. Cihlar, J. Goodison, and B. Margolis, H., 1993, BOREAS: Boreal Ecosystem-Atmosphere Study, IEEE Geoscience Remote Sensing Society Newsletter, March, 9-17.

Harris, L.D., 1984, *The Fragmented Forest: Island Biogeographic Theory and the Preservation of Biotic Diversity*. Chicago, University of Chicago Press.

Helldén, U., 1984, Remote sensing for drought impact assessment - A study of land transformation in Kordofan, Sudan. *Advanced Space Resources*, **4**, 165-168.

Helldén, U., 1988, Desertification monitoring; is the desert encroaching? *Desertifcation Control Bulletin*, **17**, 8-12.

Henderson-Sellers, A., and Pitman, A.J., 1992, Land-surface schemes for future climate models: specification, aggregation, and heterogeneity. *J. Geophys. Res.*, **97**, 2687-2696.

Hielkema, J.U., 1990 Operational environmental satellite remote sensing for food security and locust control by FAO. The ARTEMIS and DIANA systems. In: Proceedings of the ISPRS, Session: Global Monitoring TP-1, 18 Sept 1990, Victoria, B.C., Canada.

Hielkema, J.U., J. Roffey, and C.J. Tucker, 1986, Assessment of Ecological Conditions Associated with the 1980/81 Desert Locust Plague Upsurge in West Africa Using Environmental Satellite data, *Int. J. Remote Sensing*, **7**, 1609-1622.

Holben, C.J., 1986 Characteristics of maximum-value composite images for temporal AVHRR data. *Int. J. Remote Sens.*, **7**, 1435-1445.

INPE, 1992, Deforestation in Brazilian Amazonia (Instituto Nacional de Pesquisas Espaciais, Sao Jose dos Campos, Brazil).

IPCC, Intergovernmental Panel on Climate Change, 1995, The IPCC Second Assessment Synthesis of Scientific-Technical Information Relevant to Interpreting Article 2 of the UN Framework Convention on Climate Change. Cambridge, Cambridge University Press.

James, M. E. and Kalluri, S. N. V., 1994. The Pathfinder AVHRR land data set: An improved coarse resolution data set for terrestrial monitoring. *International Journal of Remote Sensing*, **15**, 3347-3363.

Jones, P.D., *et al.*, (ed.), 1985, Grid point surface air temperature dataset for the Northern Hemisphere. U.S. Dept of Energy, Washington D.C., 251 pp.

Justice, C.O., and Dowty, P., (ed.), 1994, IGBP-DIS Satellite Fire Detection Algorithm Workshop Technical Report. IGBP-DIS Working Paper, 9, 88 pp., February 1993, NASA/GSFC, Greenbelt, Maryland.

Justice, C.O. (ed.), 1986, Monitoring the grasslands of semi-arid Africa using NOAA-AVHRR data. *Int. J. Remote Sens.*, **7**, 1383-1622.

Keeling C.D., Chin, J.F.S. and Whorf, T.P., 1996, Increased activity of northern vegetation inferred from atmospheric CO2 measurements, *Nature,* **382**, 146-149.

Kimes, D.S., Holben, B.N. Tucker, C.J. and Newcomb, W.W., 1984, Optimal Directional View Angles for Remote-Sensing Missions. *Int. J. Rem. Sens.*, **5**, 887-908.

Koffi, B., Grégoire, J.-M., and Eva, H., 1996, Satellite Monitoring of Vegetation Fires on a Multi-Annual Basis and at Continental Scale, in Africa. In: *Biomass Burning and Global Change*, Levine, J. S. (ed.), Cambridge, MA, MIT Press.

Kuchler, 1983, World map of natural vegetation. Goode's World Atlas, 16th edition. , New York, Rand McNally.

Lamprey, H. F., 1975, Report on the desert encroachment reconnaissance in northern Sudan, 21 October to 10 November 1975. UNESCO/UNEP, Nairobi. Republished in Desertification Control Bulletin, **17**, 1-7.

Langaas, S., 1995, A Critical Review of Sub-Resolution Fire Detection Techniques and Principles Using Thermal Satellite Data, Ph. D. Dissertation, Department of Geography, University of Oslo, Norway.

Linthicum K.J., Anyamba, A. and Tucker, C.J., 1999, Climate and satellite indicators to forecast Rift Valley Fever epidemics in Kenya, *Science*, **285**, 397-400.

Los, S.O., 1993, Calibration adjustment of the NOAA AVHRR normalized difference vegetation index without recourse to component channels, *Int. J. Rem. Sens.,* **14**, 1907-1917.

Los, S.O., 1998a, Linkages Between Global Vegetation and Climate: An analysis based on NOAA Advanced Very High Resolution Radiometer Data. NASA Report GSFC/CR-1998-206852, NASA-CASI, Hanover Maryland. 199 pp

Los, S.O., 1998b, Estimation of the ratio of sensor degradation between NOAA-AVHRR channels 1 and 2 from monthly NDVI composites. *IEEE Transactions on Geoscience and Remote Sensing,* **36**, 206-213.

Los, S.O., Collatz, G.J. Sellers, P.J. Malmström C.M., Pollack, N.H. DeFries, R.S. Bounoua, L..Parris, M.T Tucker, C.J. and Dazlich, D.A, 2000, A global 9-year biophysical land-surface data set from NOAA AVHRR data. *J Hydrometeor.,* **1**, 183-199..

Los, S.O., G.J. Collatz, P.J. Sellers, L. Bounoua, and C.J. Tucker, 2001, Interannual variation in global vegetation, precipitation, land-surface temperature and sea-surface temperatuet al hese areas. *J. Climate* **14**, 1535-1549.

Loveland, T.R., Merchant, J.W. Ohlen, D.O. and Brown, J.F., 1991, Development of a land cover based characteristics database for the conterminous U.S., *Photogram. Eng. Remote Sens.*, **57**, 1453-1463.

MacDonald, R.B., and Hall, F.G. 1980, Global crop forecasting, *Science*, **208**, 670-674.

Malingreau, J. P., Stevens, G., and Fellows, L., 1985, Remote sensing of forest fires: Kalimantan and North Borneo in 1982-83. *Ambio*, **14**, 314-321.

Malmström CM, Thompson, M.V., Juday, G.P., Los, S.O., Randerson, J.T. and Field, C.B., 1997, Interannual variation in global-scale net primary production: Testing model estimates. *Global Biogeochemical Cycles*, **11**, 367-392.

Mann, M.E., and Park, J., 1996, Joint spatio-temporal modes of surface temperature and sea level pressure variability in the northern hemisphere during the last century, *J. Climate*, **9**, 2173-2162.

Matson, M., and Dozier, J., 1981, Identification of sub resolution High Temperature Sources Using a Thermal IR Sensor. *Photogrammetric Engineering and Remote Sensing*, **47**, 1311 - 1318.

Matthews, E., 1983, Global Vegetation and Land Use: New High-Resolution Data Bases for Climate Studies. *J. Climate App. Meteor.*, **22**, 474-487.

Menzel A, and Fabian, P., 1999, Growing season extended in Europe, *Nature*, **397**, 659-659

Monteith, J. L. and Unsworth, M. H., 1990, Principles of Environmental Physics, 2nd edition, London, Edward Arnold.

Myers, N., 1991, Tropical forests - Present status and future outlook, *Climatic Change,* **19**, 3-32.

Myneni, R.B., Keeling, C.D., Tucker, C.J., Asrar, G., and Nemani R.R., 1997, Increased plant growth in the northern high latitudes from 1981 to 1991, *Nature*, **386**, 698-702.

Myneni R.B., Los S.O., Tucker C.J., 1995, Satellite-based identification of linked vegetation index and sea surface temperature anomaly areas from 1982-1990 for Africa, Australia and South America, *Geophys Res. Let.*, **23**, 729-732.

Nicholls, N., 1991, Teleconnections and health. In: M.H. Glantz, Katz, R.W. and Nicholls, N. (ed.), pp. 493-510, New York, Cambridge University Press.

Nicholson, S. E., Kim, J. and Hoopingarner, J., 1988, Atlas of African Rainfall and its Interannual Variability. Talahassee, Florida, Department of Meteorology, Florida State University.

Olson, J.S., Watts J. and Allison, L., 1983, Carbon in live vegetation of major world ecosystems. Report No. W-7405-ENG-26, U.S. Dept. of Energy, Oak Ridge National Laboratory.

Olsson, L., 1985, Desertification or climate? Investigation regarding the relationship between land degradation and climate in the central Sudan. PhD Thesis. Lunds Studies in Geography XCVIII, Department of Geography, University of Lund, Lund, Sweden.

Potter, C.S., Randerson, J.T. and Field, C.B., 1993, Terrestrial ecosystem production: a process model based on global satellite data. *Global Biogeochem. Cycles*, **7**, 811-841.

Prince, S. D., 1991, Satellite remote sensing of primary production: comparison of results for Sahelian grasslands. *Int. J. Remote Sens.*, **12**, 1301-1311.

Prince, S.D., and Justice, C.O. (ed.), 1991, Coarse resolution remote sensing of the Sahelian environment. *Int. J. Remote Sens.*, **12**, 113-1421

Prince, S.D., Kerr, Y.H., Goutorbe, J.–P., Lebel, T., Tinga, A., Bessemoulin, P., Brouwer, J., Dolman, A.J., Engman, E.T., Gash, J.H.C., Hoepffner, M., Kabat, P., Monteny, B.F., Said, F., Sellers, P., and Wallace, J. 1995, The Hydrologic Atmospheric Pilot Experiment in the Sahel (HAPEX–Sahel). *Remote Sensing of Environment.* **51**, 215–234.

Qi, J., Chehbouni, A. Huete, A.R. Kerr, Y.H. and Sorooshian, S., 1994, A modified soil adjusted vegetation index. *Remote Sens. Environ.*, **48**, 119-126.

Randerson, J.T., Field, C.B., Fung, I., and Tans, P., 1999, Increases in early season ecosystem uptake explain changes in the seasonal cycle of atmospheric CO2 at high northern latitudes. *Geophys. Res. Let.*, **26**, 2765.

Rao, C.R. and Chen, J., 1994, Post-launch calibration of the visible and infrared channels of the advanced very high resolution radiometer on NOAA-7, -9, and 11 spacecraft. NOAA Technical Report NESDIS-78, National Oceanic and Atmospheric Administration, Washington DC 20233, 1994.

Reynolds, R.W., and Marsico, D.C., 1993, An improved real-time global sea-surface temperature analysis, *J. Climate*, **6**, 114-119.

Ropelewski, C.F., and Halpert , M.S., 1987, Global and regional scale precipitation patterns associated with the El Niño - Southern Oscillation, *Monthly Weather Rev*, **115**, 1606-1626.

Ropelewski, C.F., and Halpert, M.S., 1989, Precipitation patterns associated with the high index phase of the Southern Oscillation. *Journal of Climate*, **2**, 268-283.

Rosborough, G.W., Baldwin, D.G. , and Emery, W.J., 1994, Precise AVHRR image navigation, *IEEE Trans. Geosci. Remote Sens.*, **32**, 654-657.

Roujean, J.L., Leroy, M. Deschamps, P.Y., 1992, A bidirectional reflectance model of the earth's surface for the correction of remote sensing data, *J. Geophys. Res.-Atmos.*, **18**, 20455-20468.

Sellers, P.J., and Hall, F.G., 1992, FIFE in 1992: Results, scientific gains, and future research directions. *J. Geophys. Res.*, **97**, 19,091-19,109.

Sellers, P.J., Bounoua, L., Collatz, G.J., Randall, D.A., Dazlich, D.A., Los, S. O., Berry, J.A., Fung, I., Tucker, C.J., Field, C.B. and Jensen, T.G, 1996a, Comparison of radiative and physiological effects of doubled atmospheric CO_2 on climate. *Science*, **271**, 1402-1406.

Sellers, P.J., Los, S.O., Tucker C. J., Justice C.O., Dazlich, D.A., Collatz, G.J. and Randall, D.A., 1996b, A revised landsurface parameterization (SiB2) for GCMs. Part 2: The generation of global fields of terrestrial biophysical parameters from satellite data, *J. Climate*, **9**, 706-737.

Sellers, P.J., Randall, D.A., Collatz, G.J., Berry, J.A., Field, C.B., Dazlich D.A., Zhang, C. and Bounoua, L., 1996c, A revised land-surface parameterization (SiB2) for GCMs. Part 1: Model Formulation. *J. Climate*, **9**, 676-705.

Skidmore, A.K., 1988. Predicting bushfire activity in Australia from El Niño/Southern Oscillation events. *Australian Forestry,* **50**, 231-235.

Skidmore, A.K., Bijker, W., Schmidt, K., Kumar, L., 1998, Use of Remote Sensing and GIS For Sustainable Land Management. *ITC Journal* **1997(3/4)**,302-315.

Skole D. and Tucker, C.J., 1993, Tropical deforestation and habitat fragmentation in the Amazon - Satellite data from 1978 to 1988, *Science*, **260**, 1905-1910.

Stowe, L.L., McClain, E.P., Carey, R.M., Pellegrino, P.P., Gutman, G.G., Davis, P., Long, C. and Hart, S., 1991, Global distribution of cloud cover derived from NOAA/AVHRR operational satellite data. *Adv. Space Res.*, **3**, 51-54.

Tanre D., Holben, B.N. and Kaufman, Y.J., 1992, atmospheric correction algorithms for NOAA-AVHRR products: Theory and application. *IEEE Trans. Geosci. Remote Sens.*, **30**, 231-248

Tans, P.P., Fung, I.Y. Takahashi, T., 1990, Observational contraints on the global atmospheric CO_2 budget, *Science*, **247**, 1431-1438.

Tarpley, J.D., Schneider, S.R., and Money, R.L., 1984, Global vegetation indices from the NOAA-7 meteorological satellite. *J. Climate Applied Meteor.*, **23**, 491-494.

Tucker, C. J., 1979, Red and photographic infrared linear combinations monitoring vegetation. *Remote Sens. Environ.*, **8**, 127-150.

Tucker, C.J., 1996, History of the use of AVHRR data for land applications. In: Advances in the Use of NOAA AVHRR Data for Land Applications. In: G. D'Souza *et al.* (ed.). ECSC, EEC, EAEC, Brussels and Luxembourg, pp. 1-19.

Tucker R.P. and J.F. Richards, 1983, Global Deforestation and the Nineteenth Century World Economy. , Durham, NC, Duke University Press.

Tucker, C.J., Holben, B.N. Elgin, J.H. and McMurtrey, J.E., 1981, Remote sensing of total dry matter accumulation in winter wheat. *Remote Sens. Environ.,* **11**, 171-189.

Tucker, C.J., VanPraet, C.L., Boerwinkel, E. and Gaston, A., 1983, Satellite remote sensing of total dry accumulation in the Senegalese Sahel. *Remote Sens. Environ.*, **13**, 461-474.

Tucker, C.J., Gatlin, A. and Schneider, S.R., 1984, Monitoring vegetation in the Nile Delta with NOAA-6 and NOAA-7 AVHRR. *Photogrammetric Engineering Remote Sens.*, **50**, 53.

Tucker, C.J., Townshend, J.R.G. and Goff, T.E., 1985a, African land-cover classification using satellite data. *Science*, **227**, 369-375.

Tucker, C.J., VanPraet, C.L. Sharman, M.J. and van Ittersum, G., 1985b, Satellite remote sensing of total herbaceous biomass production in the Senegalese Sahel: 1980-1984. *Remote Sens. Environ.*, **17**, 233-249.

Tucker, C.J., Fung, I.Y. Keeling, C.D. and Gammon, R.H., 1986a, The relationship of global green leaf biomass to atmospheric CO_2 concentrations. *Nature*, **319**, 159-199.

Tucker, C.J., Justice, C. O., and Prince, S. D., 1986b, Monitoring the grasslands of the Sahel 1984-1985. *Int. J. Remote Sens.*, **7**, 1571-1582.

Tucker C.J, and P.J. Sellers, 1986, Satellite Remote Sensing of Primary Production, *Int. J. Remote Sens.*, 1986, **7**, 1395-1416.

Tucker, C.J., Dregne, H.E. and Newcomb, W.W., 1994, AVHRR data sets for determination of desert spatial extent, *Int. J. Remote Sens.*, **15**, 3547-3565.

Tucker, C.J., H.E. Dregne, W.W Newcomb, 1991, Expansion and contraction of the Sahara desert from 1980-1990, *Science*, **253**, 299-301.

UNCOD, 1977, Desertification: Its Causes and Consequences. Secretariat of United Nations Conference on Desertification. Nairobi, Kenya, Pergamon Press.

Vermote, E.F., and Kaufman, Y.J., 1995, Absolute calibration of AVHRR visible and near-infrared channels using ocean and cloud views. *Int. J. Rem. Sens.*, **16**, 2317-2340.

Vermote, E.F., Tanre, D., Seuze, J.L., Herman M. and Morcrette, J.J., 1995, 6S user guide version 1. College Park, Maryland, Department of geography, University of Maryland,.

Vermote, E., El Saleous, N., Kaufman, Y. J. and Dutton, E., 1997. Data pre-processing: Stratospheric aerosol perturbing effect on the remote sensing of vegetation: Correction method for the composite NDVI after the Pinatubo Eruption. *Remote Sensing Reviews*, **15**. 7-21.

White, F., 1983, The vegetation of Africa. Natural Resources Research vol. XX, UNESCO, Paris.

Wilson, M.F., and Henderson-Sellers, A., 1985, A global archive of land cover and soils data for use in general circulation models. *J. Climatology*, **5**, 119-143.

Wilson, E.O., 1988. In: Wilson E.O. and Peters, F.M. (ed.)., *Biodiversity,* National Academy Press, Washington, DC, 1988.

Vegetation mapping and monitoring

Curtis E. Woodcock, Scott A. Macomber and Lalit Kumar

6.1 INTRODUCTION

The nature and properties of vegetation are fundamental attributes of landscapes. The nature of the vegetation in an area is determined by a complex combination of effects related to climate, soils, history, fire and human influences which can date back several millenia in some locations. People have been interested in understanding the distribution of vegetation types since the times of Theophrastus, with significant contributions coming from such noted historical figures as Alexander von Humboldt and Lord Alfred Wallace. When viewed from this historical perspective, vegetation mapping has a long history which includes a variety of contexts and a wide range of geographic scales.

From a more modern perspective, one common distinction in vegetation mapping separates attempts to map 'potential' and 'actual' vegetation. Maps of potential vegetation attempt to determine what the vegetation type would be in the absence of human influences (Box 1981; Kuchler and Zonneveld 1988).

Maps of 'actual' vegetation attempt to characterize the vegetation as it exists in an area. Different vegetation maps emphasize different attributes of the vegetation. Some are floristic in orientation and focus on taxonomic differences between places. Others are focussed on more structural attributes of the vegetation, emphasizing the basic lifeforms of the vegetation and the size and density of cover. The characteristics emphasized in vegetation maps and their scale are typically dependent on the needs and interests of the users of the maps. At one end of the spectrum are global vegetation maps which are often used to study the relationship between vegetation types and climate (Köppen 1931; Olson et al. 1983; also see Chapter 4).

More local scales of vegetation maps are often made to serve the needs of local land management. Vegetation can be viewed in a myriad of ways from the land management perspective, including as: a source of food and/or fiber; habitat for wildlife; protector of soils; a recreational resource; a regulator of the interactions between the surface and the atmosphere with respect to heat, gases, and moisture; or simply as a fundamental attribute and descriptor of landscapes. Thus, vegetation is fundamental to many environmental processes and as a result plays a central role for the focus of this book, or the use of GIS for environmental modelling.

The goal of this chapter is to provide some history and context to recent and current efforts to map and monitor vegetation, while providing some indications regarding the way vegetation maps are used in environmental modelling. Remote sensing has revolutionized vegetation mapping, as the synoptic perspective is ideal for mapping landscape attributes. The focus here will be on the use of satellite remote sensing for vegetation mapping and monitoring, and the discussion attempts to characterize the recent innovations and ongoing areas of active research. Please note that this chapter focuses on vegetation maps at local to regional scales. Discussion of the use of satellite imagery for continental to global scales is included in Chapters 4 and 5.

6.2 VEGETATION MAPPING

6.2.1 Historical overview

The first vegetation maps made with the help of remote sensing were based on the visual interpretation of aerial photographs. The basic mapping scenario involves delineation of homogeneous patches, or stands of vegetation, for which labels are provided concerning the properties of the vegetation within the polygon. Typical vegetation properties included are the overall lifeform of the vegetation, dominant species, height and density of the vegetation, and the presence and nature of understory vegetation. Some of these properties are measured using photogrammetric methods, such as vegetation height measurements using a parallax bar (Lillesand and Kiefer 2000). Other vegetation properties are inferred from the tone, color, shape, texture, pattern, site, context and association observed in the aerial photograph (Estes *et al.* 1983) based on the knowledge of the interpreter and augmented with field visits to the area being mapped.

    With the advent of the Landsat programme in 1972, there was an immediate interest in the potential for mapping vegetation over larger areas in a more efficient manner than traditional air photo interpretation. The primary initial advantages derived from the digital format of the imagerywhich made it possible to use computers to do automated interpretation. The use of computers for analysis held great promise for reducing time and effort in vegetation mapping. Another immediate savings resulted from the digital format and geometric fidelity of the data which greatly facilitated integration of the resulting vegetation maps into GIS.

    Vegetation mapping from satellite imagery has been dominated by use of data from the reflective wavelengths of the solar spectrum, primarily the visible, near-infrared and mid-infrared. The initial sensor used for vegetation mapping was the Landsat Multispectral Scanner (MSS) which has four broad spectral bands in the visible and near-infrared wavelengths. Landsat 4 included a new sensor called the Thematic Mapper, which has 6 reflective bands with 30 m spatial resolution and a thermal band. The SPOT HRV provides finer spatial resolution (20 m) than Landsat TM but fewer spectral bands and the images cover a smaller area (see Chapter 3 for a review of sensor characteristics). Landsat TM and SPOT HRV have been the most commonly used sensors for vegetation mapping and monitoring.

    The strong reliance of air photo interpretation on the skill and experience of the interpreter is both the strength and weakness of this approach to vegetation mapping. In general, digital analysis of satellite imagery cannot match the quality of vegetation maps derived from outstanding air photo interpretation. Many vegetation maps are still made via air photo interpretation, particularly for areas small enough that the economies of scale associated with digital analysis of satellite imagery are unimportant, or where the requirements for spatial detail or accuracy of the vegetation maps are beyond those achievable with satellite remote sensing. Vegetation mapping was one of the first uses of satellite remote sensing imagery and has been one of the most common ever since (see for example Hoffer and Staff 1975).

There have been many approaches and developments involving vegetation mapping from satellite remote sensing. The discussion below is organized by the information sources exploited to map vegetation, presented in roughly the order in which they were pursued and developed.

6.2.2 Multispectral data and image classification

The first and most common approach used to map vegetation from satellite imagery is the use of multispectral data in image classification. In this approach patterns of spectral reflectance, or 'spectral signatures', are associated with different vegetation types. In the image classification step, each pixel in the image data is assigned to a particular vegetation type, resulting in a map. This paradigm has used data primarily from the solar reflective wavelengths, but other kinds of data were later included, such as texture data or other kinds of map data such as topographic variables.

Image classification algorithms can be sorted into those which are 'supervised' or 'unsupervised'. The supervised classification approaches require training sites as input prior to the image classification step which are used to characterize the spectral signatures of the vegetation types. Initially, parametric statistical classifiers such as maximum likelihood dominated (Swain and Davis 1978).

Unsupervised classification proceeds by allowing the computer to define spectral clusters of pixels, or groups of pixels in the image with similar spectral properties. Each pixel is then assigned to one such cluster. User input is necessary to associate vegetation types with spectral clusters. The primary difference between the supervised and unsupervised approaches is the timing of the user input relative to the classification step. When the input is provided ahead of classification, the approach is said to be supervized. Unsupervised approaches require user input after the classification step.

The maximum likelihood classifier assumes that the spectral signatures of vegetation types are distributed in a multivariate normal fashion, which is often not true. Vegetation classes often exhibit multimodal or non-normal shapes in their distributions, which is the result of the inherently complex nature of remote sensing images in the optical domain. Many factors influence the reflectance from vegetation canopies, some diagnostic of the vegetation types of interest in the mapping process and others unrelated. The vegetation factors known to influence the spectral reflectance of vegetation canopies include the overall life form of the vegetation, leaf properties (leaf area and leaf angle distribution and spectral reflectance properties), vegetation height or tree size, the fractional cover of vegetation, and the health and water content of leaves. In addition, the soil colour and wetness contribute to the spectral response at any given location in the image. The net effect is that the same vegetation type may have many spectral manifestations in the image, which significantly complicates the image classification process.

There have been many approaches proposed and tested to attempt to accommodate the complex nature of spectral signatures of vegetation types in the classification process. Unsupervised image classification can be formulated to include many more spectral clusters than the number of intended vegetation types, thus allowing many spectral clusters within each vegetation type. In supervised classification, many training sites can be used for single vegetation types, with individual training sites or small groups of training sites used separately through the image classification step to identify subpopulations of the intended vegetation types. These subpopulations can then be merged after the classification step to produce the final map.

There has also been significant innovations in image classifiers driven by the problems posed by vegetation mapping. For example, Skidmore and Turner (1988) developed nonparametric classification algorithms to accommodate the problems associated with non-normal distributions. More recently algorithms based on decision trees and artificial neural networks are proving to be more effective than traditional methods (Foody *et al.* 1995, Friedl and Brodley 1997, Carpenter *et al.* 1997). It should be noted that not all investigators are finding improvements with these algorithms (Skidmore *et al.* 1997), and they often are more difficult to implement as they require more training data and are not available in common image processing packages. The main strength of these algorithms with respect to vegetation mapping is to allow association of many spectral patterns to single vegetation types using a supervized approach without requiring separation of the training data into subpopulations.

One issue that confronts the use of digital satellite images for mapping vegetation concerns scale, or the relationship between the size of individual pixels and the desired scale of the resulting map. Frequently, the pixels in satellite images are too small to be classified individually in the final map. For example, at map scales common for use in local land management, such as 1:25,000, minimum mapping units are typically on the order of 1–2 hectares, or 25–50 pixels in a SPOT HRV image or 11–22 pixels in a Landsat TM image. This issue remains one of active research, and several approaches exist for this situation, including: filtering of the images resulting from per-pixel classifiers (Kim 1996); using spatial or contextual information in the classification process (Kettig and Landgrebe 1976; Stuckens *et al.* 2000), and the segmentation of images into polygons in a step independent of image classification (Woodcock and Harward 1992).

6.2.3 Vegetation mapping, ancillary data and GIS

The relationship between vegetation mapping and GIS is mutually beneficial. On the one hand, vegetation maps are used extensively within GIS for the purposes of environmental modelling, as illustrated in many ways in this book. However, the integration of other kinds of map data with remote sensing images through the use of GIS has greatly improved the vegetation mapping process. It is this dimension of the relationship between vegetation mapping and GIS which is emphasized in this section.

Vegetation mapping based solely on image classification of multispectral data is limited with respect to the vegetation attributes that can be provided in a reliable manner. Particularly apparent in this regard is the difficulty of mapping vegetation at the level of detail of individual plant species. This problem arises because many species often have overlapping spectral signatures which makes their identification impossible or of poor accuracy.

The use of topographic data to improve or augment maps made using satellite imagery dates from some of the earliest attempts to use satellite remote sensing to make vegetation maps (Hoffer and staff 1975; Strahler *et al.* 1978). The primary intent of the use of topographic data was to capture the influence of climate on species distributions, with topographic variables of elevation, aspect and slope being used as surrogates for temperature and moisture conditions. Such approaches have proven highly successful and are used frequently in vegetation mapping efforts (Skidmore 1989; Woodcock *et al.* 1994).

The detailed example given below helps illustrate the ways in which ancillary data are being integrated with remote sensing in vegetation mapping efforts.

6.2.3.1 *Modelling example: mapping* Eucalyptus *species distribution using solar radiation data*

Introduction

A number of response models have been developed to investigate the relationships between different environmental factors and the distribution of forest species (e.g. McColl 1969; Austin *et al.* 1984; Austin *et al.* 1990; Moore *et al.* 1991). These models have included environmental variables such as nutrient availability, rainfall, temperature (Moore *et al.* 1991), topographic position (Austin *et al.* 1983; Austin *et al.* 1994), elevation, aspect, exposure to wind (Mosley 1989), slope position (Twery *et al.* 1991), soil structure (Florence 1981) and soil nutrients (Turner *et al.* 1978).

Some of these models have used a solar radiation index for vegetation mapping (Kirkpatrick and Nunez 1980; Austin *et al.* 1983; Moore *et al.* 1991; Ryan *et al.* 1995), with Kirkpatrick and Nunez (1980) reporting a strong correlation between solar radiation and the distribution of several species of eucalyptus along a single transect in the Risdon Hills in Tasmania. These models have calculated solar radiation over individual field plots through field measured parameters using the method suggested by Fleming (1971) or have used radiation measuring devices, such as pyranometers (Kirkpatrick and Nunez 1980). While solar radiation data collected in the field are generally the most reliable, it is very difficult to extrapolate these data to other sites or over a large area, especially in mountainous areas where solar radiation is strongly influenced by terrain. Such data, based on point samples measured in the field, are not suited to spatial modelling in a GIS.

Solar radiation indices based solely on slope aspect and slope gradient are crude estimates as they do not take into account shading by adjacent terrain. While such a method may be acceptable in flat areas, it will not work adequately in hilly regions where shading by topographic features can account for large differences in the radiation received at a site (Kumar *et al.* 1997). Simulations in a mountainous area by Hetrick *et al.* (1993) showed that topographic shading was more important than surface orientation. Simulations at different field sites have shown that when shading by topographic features is included, approximately 30 per cent of grid cells had their total radiation reduced by 10 per cent.

Previous research (e.g. Austin *et al.* 1984) showed the distribution of vegetation responding directly to environmental factors including temperature, moisture regime and nutrient availability; and since temperature and moisture regime may be linked to solar radiation (Ahrens 1982), it is hypothesized that the distribution of vegetation should be related to solar radiation. The aim of this study was to confirm whether the distribution of *Eucalyptus* species are related to differences in solar radiation incident at a grid cell.

Solar radiation data

Solar radiation for the study area was calculated using the method proposed by Kumar *et al.* (1997). In brief, solar radiation received at a site is dependent on the azimuth and elevation of the sun, surface gradient (slope) and orientation (aspect), as well as position relative to neighbouring surfaces. Variables such as solar azimuth and solar elevation angles change continuously throughout the day and so they have to be calculated every time the intensity of solar radiation is computed. Another important factor that needs to be calculated instantaneously is shading by topographic features. In contrast, solar declination may be calculated daily, as it varies more gradually.

Aspect and slope may be easily calculated from a DEM. The additional inputs required to derive solar radiation are the latitude of the site and the Julian date (note that if calculations are required for more than one day, the start and end day are needed). For integrating the total radiation over a period (i.e. days, weeks or months), the repeat period between the instantaneous calculation of solar flux must be specified by the user. While it would be ideal to have a very short time interval to obtain accurate results, this is not always feasible because of constraints such as computational expense and the availability of a fast computer. The time interval chosen can be larger for flat terrain but has to be smaller for mountainous regions as shadowing effects will be prominent in such environments (Kumar *et al.* 1997). The radiation flux is calculated at the mid-point of each time interval to reduce shadowing effects.

Due to the forest having a natural mix of species in each plot, the data were pre-processed before analyzing the relationship between species distribution and solar radiation. In order to generate an index of solar radiation adjusted for the species composition of the plot, the radiation values for each plot were normalized according to the number of trees of each species present. Thus, for a particular species (such as *Eucalyptus sieberi*), a 'weighted mean' of the radiation values was calculated by multiplying the radiation value of each plot by the number of trees of *Eucalyptus sieberi* in the plot, adding up these values, and then dividing by the total number of trees of *E.sieberi* in all the plots. The 'weighted mean' radiation value therefore represents the overall radiation zone in which species are located. This index emphasized the plots in which a particular species is located, as the frequency of occurrence gives an indication of the environment in which those species are located. If the plot data are not normalized by the number of trees, then a plot with say 100 trees of *E.sieberi* has the same weight as a plot which has only one tree of *E.seiberi*, and both will contribute equally to the analysis. The decision to normalize is based on the observation that the forest structure is composed of 'old growth' that is dominated by an overstorey of large and medium sized trees.

A research hypothesis that the 'mean weighted' radiation values differed between species was tested using the F-test. Stated formally, the null hypothesis is that the mean solar radiation for different species are equal, that is:

Ho: $\mu_{species\ 1} = \mu_{species\ 2}$

while the alternative hypothesis is that there is a difference in mean solar radiation between species 1 and species 2,

H1: $\mu_{species\ 1} \neq \mu_{species\ 2}$

for $\alpha' < 0.05$.

Species groupings were investigated using multivariate techniques such as cluster analysis. Due to the large number of cases the K-Means cluster analysis algorithm, based on the nearest centroid sorting (Anderberg 1973), was used for determining cluster membership.

Results

Figure 6.1 shows the variation in short wave solar radiation across the study area for the different seasons, ranging from 3600 to 11200 $MJ/m^2/year$. Exposed sites received almost 3 times as much solar radiation as shadowed sites.

The seasonal difference in the solar radiation is large, and as expected, the summer season receives more radiation than the winter season. The mean radiation is highest in summer, followed by spring, autumn and winter respectively, with the mean winter value being 60 per cent of the summer value. Many gully sites do not receive any direct radiation during winter as adjacent ridges continually shade them.

The variance in solar radiation is least during summer and largest in winter, especially for the south, southeast and southwest aspects (Note that the study site is located in the southern hemisphere). These aspects have the largest variation on their midslopes, caused by shading when the sun is lower in the winter sky.

North facing slopes receive far more radiation than the south facing slopes, with other aspects ranging between these two extremes. While the radiation values for north, north-east and north-west aspects stay almost constant with slope, those for south, south-east and south-west fall off fairly sharply as the slope increases. At low slopes, there is little difference in the radiation values for the different aspects, with the differences between aspects being least in the summer months.

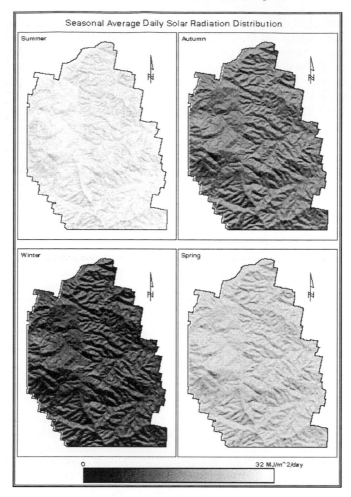

Figure 6.1: Solar radiation distribution for the different seasons.

Figure 6.2 shows that different species are located in certain radiation zones. Of the 12 species in this study, *E.smithii* and *E.consideniana* occur at sites with the lowest radiation values. *E.sieberi*, *E.globoidea* and *E.viminalis* are consistently placed at the high end of radiation values. Seasonal variation in solar radiation was also found to influence the occurrence of *Eucalyptus* species. Results of Student-Newman-Keuls Test showed that, for many of the species, the differences in the mean radiation were statistically significant. For example *E.consideniana* had insolation values which were significantly different from all species for all the seasons, except in spring and summer where *E.smithii* was an exception. Similarly *E.obliqua* was significantly different from all the species for all the seasons. Some other species that returned significance with many of the species were *E.agglomerata*, *E.bosistoana*, *E.cypellocarpa*, *E.globoidea* and *E.muellerana*.

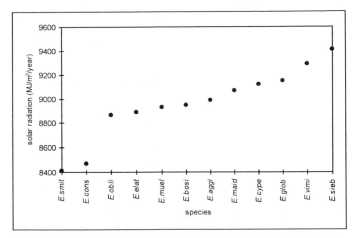

Figure 6.2: Relationship between the 'weighted mean' solar radiation and the different species.

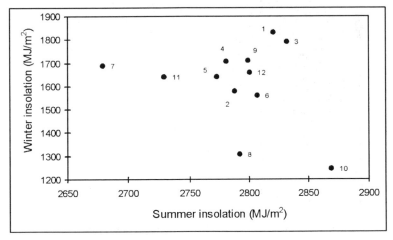

Figure 6.3: Species position in relation to their summer and winter radiation values.

[1. *E.agglomerata*, 2. *E.bosistoana*, 3. *E.consideniana*, 4. *E.cypellocarpa*, 5. *E.elata*, 6. *E.globoidea*, 7. *E.maidenii*, 8. *E.muellerana*, 9. *E.obliqua*, 10. *E.sieberi*, 11. *E.smithii*, 12. *E.viminalis*]

Another important aspect that was noted was that the confidence intervals shifted between seasons, indicating that seasonal differences in radiation may assist in characterizing species. For example *E.sieberi* had a confidence interval placed at the high end of radiation in summer and was at the upper end of the interval for *E.viminalis* but moved lower in autumn and winter seasons and, when compared to *E.viminalis*, was at the lower end of its confidence interval. Therefore, while the confidence intervals changed by seasons, much as expected, they also shifted position relative to other species by seasons (Kumar and Skidmore 2000). It is these changes in position relative to other species over different seasons that may be used to characterize species. The winter season showed the greatest variation in mean radiation and the confidence intervals were more separated. Figure 6.3 shows how species separate when winter insolation values for each species are plotted against their summer values.

Possible uses of insolation data in forestry modelling

Since many species show that their radiation regimes are significantly different from other species, this information can be utilized to delineate individual species or to find the most likely habitats of the species. The different seasonal confidence intervals of means for each species can be combined to make the selection criteria. For example, from the means tables, the conditions for the different species would be used to produce probability maps as given in Figure 6.4. Diagrams such as these can then be used for planning logging operations, habitat mapping, conservation work, etc. These figures do not confirm with 100 per cent certainty that the particular *Eucalyptus* species would be found at the mapped site, but they pinpoint the most probable sites of occurrence and hence a good starting point. The Australian Koala Foundation is already using the insolation model for the prediction of koala populations. Koalas feed on specific species of *Eucalyptus* and if the habitat of these species is mapped out then the possible locations of koalas can be predicted.

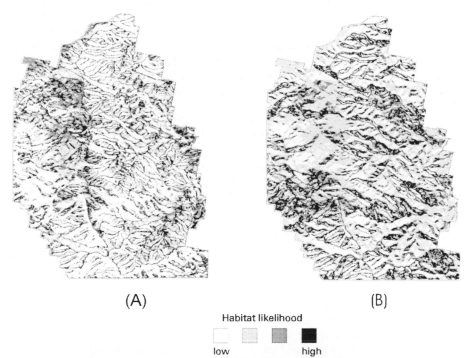

(A)   (B)

Habitat likelihood

low   high

Figure 6.4: Possible locations of *E.sieberi* (a) and *E.consideniana* (b) based on seasonal insolation data.

Similar work can be extended to forest types as well. If a large number of plots are surveyed and confidence intervals of means are produced as done for the species distribution, then a similar set of conditions can be coded. This should show the most likely locations for the different forest types.

However the main use of solar radiation in modelling vegetation distribution would be as a value added layer in other environmental models (Figure 6.5). As mentioned before, solar radiation is one of a number of environmental factors that affect species distribution. Other researchers have shown the correlation between species and a number of other environmental factors. These factors can be combined with solar insolation to model the species more effectively. In a GIS, each of these factors can be stored as a different layer and Boolean conditions can be coded to model the different species. Solar radiation modelled in this manner can also be used as an input parameter in expert systems or neural networks.

6.2.4 Use of spatial and temporal patterns

Two additional kinds of information which have been used in vegetation mapping based on satellite imagery are spatial and temporal patterns. The use of spatial patterns, or texture, is based on the long recognized value of texture in air photo interpretation for differentiating vegetation types. To use texture in vegetation mapping using satellite imagery, a new texture band is created from one of the original spectral bands. The texture band (or bands) are then combined with the original spectral bands in the image classification process, increasing the number of input bands. It represents an attempt to exploit in automated image classification one kind of information which contributes greatly to visual interpretation of air photos. Several studies have shown the use of texture data to improve vegetation maps derived from satellite imagery (Franklin *et al.* 1986; Franklin and Peddle 1989; Jakubauskas 1997).

Temporal patterns, or the change in reflectance properties over time, have been used extensively for mapping vegetation at continental to global scales using NOAA AVHRR imagery (See Chapter 3). The NOAA imagery has coarse spatial resolution but high temporal resolution, so phenological patterns of vegetation can be captured using AVHRR imagery (DeFries and Townshend 1994). The basic approach is to use multiple dates of imagery as input bands to image classification procedures. More recently, several investigators have found that use of multitemporal imagery can improve vegetation mapping at local to regional scales using imagery such as Landsat TM or SPOT HRV (Wolter, *et al.* 1995; Mickelson *et al.* 1998). The use of multitemporal imagery appears most promising in environments with mixes of evergreen and deciduous species.

Model output as an input to other more sophisticated models

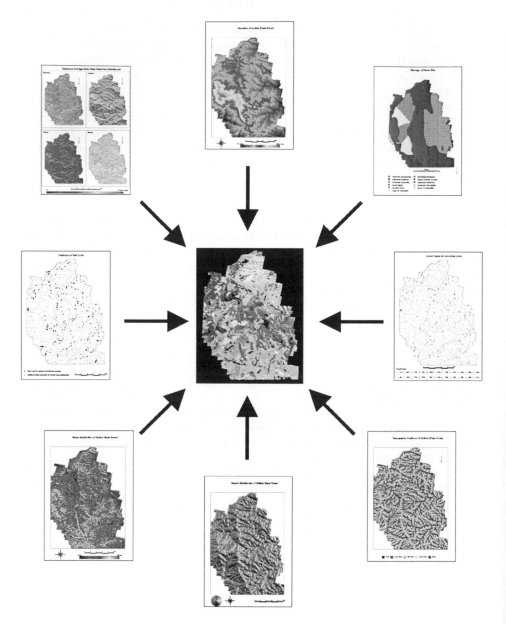

Figure 6.5: Solar radiation data form the GIS model as input into another larger GIS model.

6.2.5 New kinds of imagery

There are a variety of new kinds of imagery being used, or at least experimented with, for vegetation mapping. They hold great potential for improving the kinds of information that can be provided about vegetation via remote sensing in the future. These new types of imagery are described in detail in Chapter 3.

6.2.5.1 Hyperspectral imagery

While data obtained from broadband sensors (such as the Landsat TM and ETM+ and SPOT HRV) have been useful in many respects for vegetation mapping, they also have their limitations. Because of their limited number of channels and wide bandwidths, a lot of the data about plant reflectance is lost. Most natural objects have characteristic features in their spectral signatures which distinguish them from others and many of these characteristic features occur in very narrow wavelength regions (Figure 6.6).

Hence to sense these narrow features the use of narrow band sensors is required. Broadband sensors integrate the reflectance over a wide range and so the narrow spectral features are lost or masked by other stronger features surrounding them. For this reason hyperspectral remote sensing, often with bandwidths of only 5–10 nm, offers a powerful tool for significant advancement in the understanding of the Earth and its environment. A number of these narrow-band imaging spectrometers have been discussed in Chapter 3.

Figure 6.7 shows typical spectral reflectance data of vegetation as collected by a spectrometer (GER IRIS) and a simulated model of what the resulting signal would be from Landsat TM. Notice that the hyperspectral data includes detailed spectral features characteristic of vegetation which are lost in broadband sensors. Thus hyperspectral data holds the potential for providing more detailed information about vegetation than is possible with broadband sensors. While several hyperspectral sensors are planned for future satellites, current research is based on airborne systems, which are reviewed briefly in Chapter 3. Research has shown that hyperspectral remote sensing has a lot to offer with respect to species identification (Kumar and Skidmore 1998).

Also, data from airborne imaging spectrometers have been found to yield higher quality information about vegetation health and cover than those obtained from broadband sensors (Collins *et al.* 1983, Curran *et al.* 1992, Peuelas *et al.* 1993, Carter 1994, Carter *et al.* 1996, Kraft *et al.* 1996). Gamon *et al.* (1993) used the narrow AVIRIS spectral bands to evaluate the spatial patterns of vegetation type, productivity, and physiological activity in annual grasslands and the results showed the major vegetation types and fine scale patterns not discernible from broadband data.

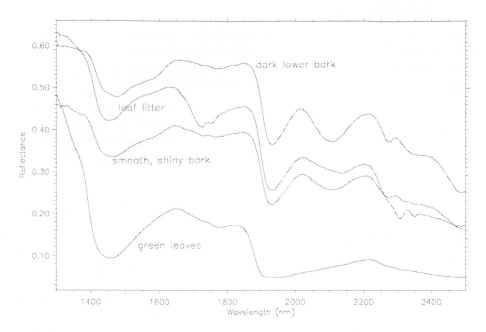

Figure 6.6: Hyperspectral data showing some spectral fine features in green leaves and different bark types in *Eucalyptus sieberi*. Note the features around 1750 nm, 2270 nm, 2300 nm and 2350 nm.

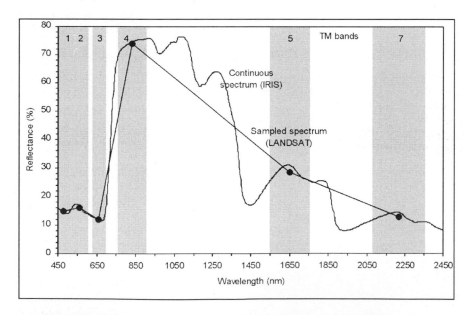

Figure 6.7: Data content of broadband (Landsat) and narrow-band (IRIS) sensors.

6.2.5.2 Radar imagery

Experiments with the use of radar for studying vegetation have demonstrated considerable promise. Research has found that radar can be sensitive to vegetation structure and biomass, particularly with multiband radar systems including lower frequencies (P- and L-band) with cross polarizations (Kasischke *et al.* 1997). There has also been progress on the use of radar imagery for general land cover mapping, which provides information on basic vegetation types. One of the main benefits of radar imagery is the independence of sensing from solar illumination, which allows for effective sensing during cloudy periods or even at night. This benefit is particularly valuable in areas characterized by high frequency of cloud cover, as is the case in many equatorial regions. Many new spaceborne radar systems have been launched recently, and a review of their characteristics has been provided in Chapter 3. The improving availability of radar imagery is destined to speed the pace with which radar imagery is adopted for vegetation mapping. The potential for combining radar and optical imagery to improve vegetation mapping is high and largely untapped at this point in time.

6.2.5.3 High spatial resolution imagery

One trend is toward the collection from satellites of imagery with high spatial resolution. Currently there is one operational satellite system, IKONOS, providing 1 m panchromatic and 4 m multispectral imagery. Other private sector systems are planned with similar spatial resolutions (see Chapter 3 for details). In anticipation of such capability, a number of airborne systems have been developed to allow for development of methods for analyzing high spatial resolution images (Franklin 1994).

Particularly in Canada there has been considerable effort devoted to learning how to use high spatial resolution imagery for vegetation mapping. The high resolution imagery contains effects associated with individual trees, and progress has been made on the problem of how to identify individual tree crowns and their size (Wulder *et al.*, 2000). Another area of active research is the estimation of tree size and cover from high resolution imagery through the analysis of observed spatial patterns in images (St Onge *et al.* 1997). The availability of satellite imagery with very high spatial resolution is destined to improve the quality of information about vegetation canopies, at least for selected areas.

6.2.6 Accuracy assessment

One important issue regarding the use of vegetation maps derived from remote sensing is accuracy. All vegetation maps contain errors, and the significance of those errors is dependent on the manner in which the vegetation maps are used. One result is that the accuracy requirements for the same map may vary between potential users of the map! Thus, careful characterization of the accuracy of vegetation maps is essential for their informed use. The most common approach used to determine the accuracy of vegetation maps is to conduct an accuracy assessment. While there are many ways to conduct an accuracy assessment, the most common is to populate an error matrix (also referred to as a confusion matrix) based on samples selected from the vegetation map. Such an approach allows estimation of the categorical nature of the errors and their overall frequency. Jensen (1996) provides a helpful discussion of the many issues and decisions involved in conducting an accuracy assessment.

Table 6.1 is an example of a hypothetical accuracy assessment for a vegetation map including four vegetation types. By convention, most confusion matrices are created with the map labels on the rows, and the reference data, or truth, down the columns. Below the confusion matrix, the Producer's and User's Accuracies are calculated for each class, as well as the overall accuracy and an accuracy estimate that removes the effect of random chance on accuracy, referred to as the Khat statistic (Skidmore 1999).

Many useful things can be learned from analysis of the accuracy assessment. The simplest statistic to derive is the overall accuracy. This is simply the sum of the diagonal elements divided by the total number of pixels (or sites) evaluated. In this case, the overall accuracy is moderate at 82 per cent, but the level of accuracy is highly variable between classes.

To better understand the variability of the accuracies of the different classes, one can also calculate the Producer's and User's accuracies. The Producer's accuracy is the number of correct elements for a class divided by the total number of pixels (or sites) given that map label (the row total). The User's accuracy is the number of correct elements divided by the total number of pixels that should truly have that label (the column total). In this context, analysts often discuss *errors of omission* and *errors of commission*. Errors of omission are those pixels which were missed by the Producer and thus are calculated as 100 minus the Producer's accuracy. By extension, errors of commission are the pixels wrongly assigned to a class and are calculated as 100 minus the User's accuracy. Thus, each error of omission from one class is also and error of commission for another class.

Table 6.1: A hypothetical result from an accuracy assessment, including a confusion matrix, and calculation of the Producer's, User's and overall accuracies (see Jensen 1996 for details).

Map Classification	Reference Data				
	A	B	C	D	Row Total
Map Class A	178	3	10	0	191
Map Class B	0	38	2	0	40
Map Class C	5	25	58	19	107
Map Class D	2	9	0	68	79
Column Total	185	75	70	87	417

Producer's Accuracy		User's Accuracy
Class A:	178/185 = 96%	178/191 = 93%
Class B:	38/75 = 51%	38/40 = 95%
Class C:	58/70 = 83%	58/107 = 54%
Class D:	68/87 = 78%	68/79 = 86%

Overall Accuracy 342/417 = 82% ;   Khat Statistic = 74%

In the example, Class A is clearly the most accurately mapped, with high Producer's and User's accuracies (96 per cent and 93 per cent respectively). In contrast, Class B has a high User's accuracy (95 per cent), meaning you can be very confident of sites identified as Class B on the map, but the Producer's accuracy is very low (51 per cent) meaning the mapping process missed about half the area that is truly class B (49 per cent errors of omission).

Class C on the other hand has a high Producer's accuracy (83 per cent). The mapping process has found most of Class C, but a site on the map identified as Class C is only correct about half the time (54 per cent User's accuracy and 46 per cent errors of commission).

Finally, Class D is moderately accurate, but examination of the confusion matrix shows that most of the problems are errors of omission. The 78 per cent Producer's accuracy means 22 per cent errors of omission. Most of the errors of omission (19/30) are mislabelled as Class C. Thus, to improve the accuracy of class D, one might begin with reevaluation of the areas mapped as class C.

6.3 MONITORING VEGETATION CHANGE

Vegetation health, condition and change through time are of great interest from a variety of perspectives. Satellite imagery, primarily due to its synoptic views of landscapes and multitemporal sensing, is well suited for monitoring vegetation health and change through time. One of the benefits of continued collection of satellite imagery by programs like Landsat and SPOT is the ability to study changes in landscapes over time, with changes in vegetation being among the most common features studied. The historical archive of satellite imagery for studying landscape change continues to grow and its duration now covers more than a quarter of a century. While this chapter highlights many ways in which this imagery is being used to study vegetation change, it is noteworthy that recently there has been a dramatic increase in studies using the archive of historical satellite imagery. This trend indicates the growing value of this archive of imagery and points to a future where remote sensing data plays a key role in our understanding of how landscapes are changing and how humans are influencing the health of vegetation.

Like vegetation mapping, imagery from the optical domain (Landsat and SPOT) have dominated efforts to monitor vegetation change. Many kinds of vegetation changes have been monitored in many different contexts and regions of the world and using a wide variety of methods. The simple taxonomy presented below emphasizes the different kinds of vegetation change being monitored with remote sensing and the numbers of images being used. The examples mentioned are far from a complete inventory of the ways remote sensing is being used in this context, but they are intended to serve as representative of the kinds of problems and range of geographic locations being studied.

6.3.1 Monitoring vegetation condition and health

Many factors influence vegetation condition and health, ranging from drought and pests to acid rain and air pollution. While it is necessary to characterize the nature of the problem and the ranges of magnitude of effects on the vegetation through field samples, it is difficult to determine the geographic extent and locations of the areas affected using conventional field methods. Remote sensing offers an alternative approach whose strengths are in spatial coverage, which when merged with field samples has been shown to be extremely helpful for monitoring vegetation health.

6.3.1.1 Single date assessments

Damage from insects is a key concern for forest health monitoring. Information on the location of damage and its severity is essential to local resource management. Williams and Nelson (1986) report on the ability to use Landsat MSS data and ratios of NIR to red reflectance to map defoliation due to insects in the hardwood forests of the north-eastern USA. Ekstrand (1990) similarly found Landsat TM data useful for monitoring levels of insect damage in Swedish conifer forests using single dates of Landsat TM imagery in the late summer.

Some of the most dramatic anthropogenic effects on vegetation health arise from air pollution and associated acid rain. Vogelmann and Rock (1988) mapped forest damage in high elevation forests of the North-eastern US using Landsat imagery. They found that a ratio of the mid IR reflectance to near infrared reflectance was diagnostic of forest damage due to acid rain/air pollution in conifer forests. Ardo *et al.* (1997) were able to map three classes of forest damage, in the form of levels of needle loss, in spruce forests in the Czech Republic using Landsat data. The forest damage classes were defined on the basis of regressions between needle loss and TM spectral data. Their study helped quantify the magnitude of deforestation and forest damage resulting from extreme acid rain/air pollution problems in this region.

6.3.1.2 Multitemporal analysis

While there has been success at times using single acquisitions of satellite imagery to map the locations of vegetation affected by such factors as insects and air pollution/acid rain, it has been much more common to use multitemporal satellite imagery to monitor vegetation health. In this approach, images from different dates for the same location are coregistered such that spectral values from the two dates can be directly compared.

In several settings, analysis of multidate images has proven effective for monitoring defoliation of forests due to insects. Muchoney and Haack (1994) used multitemporal SPOT imagery for identifying changes in hardwood forest defoliation due to gypsy moths in the eastern US. They tested a variety of methods and found the best results using image differencing and principal components analysis. In a later study, Radeloff *et al.* (1999) used spectral mixture analysis to measure the magnitude of defoliation in pine forests in north-western Wisconsin, US. Their study illustrated the importance of controlling for factors like the presence of hardwoods within pine stands on the effectiveness of defoliation monitoring. In related studies, Macomber and Woodcock (1994) and Collins and Woodcock (1994) studied drought-induced mortality in conifer forests using multitemporal Landsat imagery. The vegetation change in this case is still caused by insects, but the insects kill the trees, which is measured as a change in canopy cover or basal area. They tested a variety of methods and found that many worked well for estimating mortality within forest stands, providing an encouraging indication of the ability to detect subtle changes in canopy cover over time with multispectral satellite imagery.

Coppin and Bauer (1994) report on development of methods for operational monitoring of forest change in Minnesota, US. Their methods identify changes in forest cover due to a variety of reasons, including storm damage. They used multitemporal images and found that changes in the overall brightness and greenness of forest stands were reliable indicators of forest change. Another cause of vegetation change is air pollution, which was found to significantly damage the forests in the Kola Peninsula of northern Russia (Rigina *et al.* 1999). The air pollution was the result of the smelting industry and caused extensive damage which was monitored using satellite imagery from 1978 and 1996. Analysis of the patterns of vegetation damage indicated the influence of the surrounding mountains on the location of areas protected from damage.

6.3.2 Vegetation conversion and change

Another kind of change in vegetation of great interest is wholesale conversion of vegetation types. The most obvious example of this kind of change is deforestation, which is one of the most significant forms of land-use change occurring on Earth. Whether deforestation is due to the harvest of wood products, conversion of land to other uses such as agriculture or urban uses, the result of forest fires, or some combination of the above, monitoring of deforestation is a concern in many regions of the world. Remote sensing has been the primary tool used for monitoring deforestation. Experience has shown that deforestation is best monitored using medium resolution sensors such as Landsat and SPOT, as coarse resolution sensors such as AVHRR often produce misleading estimates of the total area deforested.

There are a variety of reasons for monitoring forest clearing, or deforestation. One reason is to understand the role of forest change in the global carbon budget, which requires data on deforestation over large areas. The best known example in this regard is the ongoing effort to monitor deforestation in Amazonia (Skole and Tucker 1993). Another reason for monitoring forest clearing is local land management. In Finland, where forests are actively managed for wood products, multitemporal Landsat TM images have been shown to be useful for providing timely information on rapid changes in forest cover (Varjo 1997). India is suffering from serious depletion of its forest cover, and remote sensing is playing a valuable role in providing information on the location and extent of forest clearing (Singh 1986; Jha and Unni 1994). Research continues with regard to how to best monitor forest clearing over large areas. Much of the initial deforestation work was based on visual interpretation of images, but recent efforts have indicated the viability of using automated analysis of multitemporal images to monitor forest change (Cohen *et al.* 1998).

6.4 CONCLUDING COMMENTS

Vegetation is a fundamental attribute of landscapes which influences a whole host of environmental processes. Mapping of vegetation via remote sensing is providing information on vegetation properties for large parts of the world in sufficient spatial detail to aid environmental modelling. Vegetation mapping at local to regional scales is currently dominated by imagery from the Landsat and SPOT satellites, but future vegetation mapping will be improved by use of hyperspectral imagery, radar imagery, and high spatial resolution imagery. Monitoring of vegetation change using remote sensing is providing an improved understanding of the health and condition of vegetation as well as rates of conversion of natural vegetation to other land uses. The value of the historical archive of satellite imagery is being repeatedly demonstrated in an increasing number of vegetation monitoring projects.

6.5 REFERENCES

Ahrens, C.D., 1982, *Meteorology Today*, St. Paul, MN, West Publishing Company.

Anderberg, M.R., 1973, *Cluster analysis for applications*. New York., Academic Press.

Aplin, P., Atkinson, P. and Curran, P., 1997, Fine spatial resolution satellite sensors for the next decade, *International Journal of Remote Sensing*, **18**, 3873–3881.

Ardo, J., Pilesjo, P. and Skidmore, A.K 1997, Neural networks, multitemporal Landsat Thematic Mapper data and topographic data to classify forest damages in the Czech Republic, *Canadian Journal of Remote Sensing*, **23**, 217–229.

Austin, M.P., Cunningham, R.B. and Fleming, M.P., 1984, New approaches to direct gradient analysis using environmental scalars and statistical curve-fitting procedures *Vegetatio*, **55**, 11–27.

Austin, M.P., Cunningham, R.B. and Good, R.B., 1983, Altitudinal distribution of several Eucalypt species in relation to other environmental factors in southern New South Wales. *Australian Journal of Ecology*, **8**, 169–180.

Austin, M.P., Nicholls, A.O., Doherty, M.D. and Meyers, J.A., 1994, Determining species, response functions to an environmental gradient by means of a ß-function. *Journal of Vegetation Science*, **5**, 215–228.

Austin, M.P., Nicholls, A.O. and Margules, C.R., 1990, Measurement of the realised qualitative niche: Environmental niches of five Eucalyptus Species. *Ecological Monographs*, **60**, 161–177.

Box, E.O., 1981, *Macroclimate and Plant Forms*. The Hague, Netherlands, Dr. W. Junk Publishers.

Carpenter, G.A., Gjaja, M.N., Gopal, S. and Woodcock, C.E., 1997, ART neural networks for remote sensing: Vegetation classification from Landsat TM and Terrain Data. *IEEE Transactions on Geoscience and Remote Sensing*, **35**, 308–325.

Carter, G.A., 1994, Ratios of leaf reflectances in narrow wavebands as indicators of plant stress. *International Journal of Remote Sensing*, **15**, 697–703.

Carter, G.A., Cibula, W.G. and Miller, R.L., 1996, Narrow-band reflectance imagery compared with thermal imagery for early detection of plant stress. *Journal of Plant Physiology*, **148**, 515–522.

Cohen, W.B., Fiorella, M. and K. Anderson, 1998, An efficient and accurate method for mapping forest clearcuts in the Pacific Northwest using Landsat imagery. *Photogrammetric Engineering and Remote Sensing*, **64**, 293–300.

Collins, W., Chang, S.H., Raines, G., Canney, F. and Ashley, R., 1983, Airborne biogeophysical mapping of hidden mineral deposits. *Economic Geology and the Bulletin of the Society of Economic Geologists*, **78**, 737–749.

Collins, J.B. and Woodcock, C.E., 1994, Change detection using the Gramm-Schmidt transformation applied to mapping forest mortality. *Remote Sensing of Environment*, **50**, 267–279.

Congalton, R.G., Oderwald, R.G. and Mead, R.A., 1983, Assessing Landsat classification accuracy using discrete multivariate analysis statistical techniques. *Photogrammetric Engineering and Remote Sensing*, **49**, 1671–1678.

Coppin, P.R. and Bauer, M.E., 1994, Processing of multitemporal Landsat TM Imagery to optimize extraction of forest cover change features. *IEEE Transactions on Geoscience and Remote Sensing*, **32**, 918–927.

Curran, P.J., Dungan, J.L., Macler, B.A., Plummer, S.E. and Peterson, D.L., 1992, Reflectance spectroscopy of fresh whole leaves for the estimation of chemical concentration. *Remote Sensing of Environment*, **39**, 153–166.

DeFries, R.S. and Townshend, J.R.G., 1994, NDVI-derived land cover classification at global scales. *International Journal of Remote Sensing*, **15**, 3567–3586.

Ekstrand, S., 1990, Detection of moderate damage on Norway Spruce using Landsat TM and digital stand data. *IEEE Transactions on Geoscience and Remote Sensing*, **28**, 685–692.

Estes, J.E., Hajic, E.J. and Tinney, L.R., 1983, Fundamentals of Image Analysis: Analysis of Visible and Thermal Infrared Data. In: *The Manual of Remote Sensing, Second Edition*, R.N. Colwell, (ed.), American Society of Photogrammetry.

Fleming, P.M., 1971, The calculation of clear day solar radiation on any surface. Paper presented at *Aust. Inst. Refrig. Air. Cond. Heating Conference*, Perth, May 1971.

Florence, R.G., 1981, The biology of the eucalypt forest. *Biology of Native Australian Plants*. J. Pate and A. McComb, (eds.), University of W.A. Press, Perth, Australia.

Foody, G.M., Mcculloch, M.B. and Yates, W.B., 1995, Classification of remotely-sensed data by an artificial neural-network – Issues related to training data characteristics. *Photogrammetric Engineering and Remote Sensing*, **61**, 391–401.

Franklin, S., 1994, Discrimination of subalpine forest species and canopy density using digital casi, SPOT LA, and Landsat TM data. *Photogrammetric Engineering and Remote Sensing*, **60**, 1233–1241.

Franklin, J., Logan, T.L., Woodcock, C.E. and Strahler, A.H., 1986, Coniferous forest classification and inventory using Landsat and digital terrain data. *IEEE Transactions on Geoscience and Remote Sensing*, **GE-24**, 39–149.

Franklin, S.E. and Peddle, D.R., 1989, Spectral texture for improved class discrimination in complex terrain. *International Journal of Remote Sensing*, **8**, 310–314.

Friedl, M.A. and Brodley, C.E., 1997, Decision tree classification of land cover from remotely sensed data. *Remote Sensing of Environment*, **61**, 399–409.

Gamon, J.A., Field, C.B., Roberts, D.A., Ustin, S.L. and Valentini, R., 1993, Functional patterns in an annual grassland during an AVIRIS overflight. *Remote Sensing of Environment*, **44**, 239–253.

Hetrick, W.A., Rich, P.M. and Weiss, S.B., 1993, Modelling isolation on complex surfaces. *Proceedings of the Thirteenth Annual ESRI User Conference* (Redlands: ESRI) 2, 447–458.

Hoffer, R.M. and Staff, 1975, *Natural resource mapping in mountainous terrain by computer-analysis of ERTS-1 satellite data, Research Bull 919*, Agricultural Experiment Station and Lab, for Applications of Remote Sensing, West Lafayette, IN. Purdue University.

Jha, C.S. and Unni, N.V.M., 1994, Digital change detection of forest conversion of a dry tropical Indian forest region. *International Journal of Remote Sensing*, **15**, 2543–2552.

Jakubauskas, M.E., 1997, Effects of forest succession on texture in Landsat Thematic Mapper imagery. *Canadian Journal of Remote Sensing*, **23**, 257–263.

Jensen, J.R., 1996, *Introductory Digital Image Processing: A Remote Sensing Perspective, Second Edition*, New Jersey, Prentice Hall, 318 pp.

Kasischke, Eric S., Melack, J.M. and Dobson, M.C., 1997, The use of imaging radars for ecological applications – A review. *Remote Sensing of Environment*, **59**, 41–156.

Kettig, R.L. and Landgrebe, D.A., 1976, Classification of multispectral iamge data by extraction and classification of homogeneous objects. *IEEE Transactions on Geoscience Electronics*, **14**, 19–26.

Kim, K.E., 1996, Adaptive majority filtering for contextual classification of remote sensing data. *International Journal of Remote Sensing*, **17**, 1083–1087.

Kirkpatrick, J.B. and Nunez, M., 1980, Vegetation-radiation relationships in mountainous terrain: Eucalypt dominated vegetation in the Risdon Hills, Tasmania. *Journal of Geography*, **7**, 197–208.

Köppen, V., 1931, *Grundriss der Klimakunde*. Berlin, Walter de Gruyter Verlag.

Kraft, M., Weigel, H., Mejer, G. and Brandes, F., 1996, Reflectance measurements of leaves for detecting visible and non-visible ozone damage to crops. *Journal of Plant Physiology*, **148**, 148–154.

Küchler, A.W. and Zonneveld, I.S., 1988, *Vegetation Mapping*. Kluwer Academic Publishers, Dordrecht, 635

Kumar, L. and Skidmore, A.K., 1998, Use of derivative spectroscopy to identify regions of differences between some Australian eucalypt species. *Proceedings 9th Australasian Remote Sensing and Photogrammetry Conference*, Sydney, Australia, July 20–24.

Kumar, L. and Skidmore, A.K., 2000, Radiation – vegetation relationships in an Eucalyptus forest. *Photogrammetric Engineering and Remote Sensing*, **66**, 193–204.

Kumar, L., Skidmore, A.K. and Knowles, E., 1997, Modelling topographic variation in solar radiation in a GIS environment. *International Journal of Geographical Information Science*, **11**, 475–497.

Lillesand, T.M. and Kiefer, R.W., 2000, *Remote Sensing and Image Interpretation, 4th edition*. New York, John Wiley and Sons, Inc., 724 pp.

Macomber, S. and Woodcock, C.E., 1994, Mapping and monitoring conifer mortality using remote sensing in the Lake Tahoe Basin. *Remote Sensing of Environment*, **50**, 255–266.

McColl, J.G., 1969, Soil-plant relationships in a Eucalyptus forest on the south coast of New South Wales. *Ecology*, **50**, 354–362.

Mickelson, John G., Jr., Civco, D.L. and Silander, J.A., Jr., 1998, Delineating forest canopy species in the Northeastern United States using multi-temporal TM imagery. *Photogrammetric Engineering and Remote Sensing*, **64**, 891–904.

Moore, D.M., Lees, B.G. and Davey, S.M., 1991, A new method for predicting vegetation distributions using decision tree analysis in a Geographic Information System. *Environmental Management*, **15**, 59–71.

Mosley, G., 1989, *Blue mountains for world heritage*. Colony Foundation for Wilderness, Sydney.

Muchoney, D.M. and Haack, B.N., 1994, Change detection for monitoring forest defoliation. *Photogrammetric Engineering and Remote Sensing*, **60**, 1243–1251.

Olson, J.S., Watts, J. and Allison, L., 1983, *Carbon in live vegetation of major world ecosystems*, W-7405-ENG-26, U.S. Department of Energy, Oak Ridge National Laboratory.

Peuelas, J., Filella, I., Biel, C., Serrano, L. and Save, R., 1993, The reflectance at the 950–970nm region as an indicator of plant water status. *International Journal of Remote Sensing*, **14**, 1887–1905.

Radeloff, V.C., Mladenoff, D.J. and Boyce, M., 1999, Detecting Jack Pine budworm defoliation using spectral mixture analysis: separating effects from determinants. *Remote Sensing of Environment*, **69**, 156–169.

Ryan, P., Coops, N., Austin, M. and Binns, D., 1995, Incorporating soil chemical data into environmental models for predicting forest species distribution in south-east NSW. *Institute of Foresters of Australia 16th Biennial Conference*, 18–21 April, Ballarat, Victoria.

Singh, A., 1986, Change detection in the tropical forest environment of northeastern India using Landsat. In *Remote Sensing and tropical land management*, Eden, M.J. and Parry, J.T. (eds), John Wiley and Sons, 237–254.

Skidmore, A.K., 1989, An expert system classifies eucalypt forest types using thematic mapper data and a digital terrain model. *Photogrammetric Engineering and Remote Sensing*, **55**, 1449–1464.

Skidmore, A.K. and Turner, B.J., 1988, Forest mapping accuracies are improved using a supervised nonparametric classifier with spot data. *Photogrammetric Engineering and Remote Sensing*, **54**, 1415–1421.

Skidmore, A.K., Turner, B.J., Brinkhof, W. and Knowles, E., 1997, Performance of a neural network: Mapping forests using GIS and remotely sensed data. *Photogrammetric Engineering and Remote Sensing*, **63**, 501–514.

Skidmore, A.K., 1999, Accuracy assessment of spatial information. In: Stein, A., van der Meer, F., and Gorte, B. (ed.). *Spatial Statistics for Remote Sensing*, Dordrecht, Kluwer Academic Publishers. pp. 197-209.

Skole, D. and Tucker, C.J., 1993, Tropical deforestation and habitat fragmentation in the Amazon: Satellite data from 1978 to 1988. *Science*, **260**, 1905–1910.

St Onge, Benoit A. and Cavalais, F., 1997, Automated forest structure mapping from high resolution imagery based on directional semivariogram estimates. *Remote Sensing of Environment*, **61**, 82–95.

Strahler, A.H., Logan, T.L. and Bryant, N.A., 1978, Improving forest cover classification accuracy from Landsat by incorporating topographic information. *Proceedings of the 12th International Symposium on Remote Sensing of Environment*, Ann Arbor, MI, 1541–1557.

Stuckens, J., Coppin, P.R. and Bauer, M.E., 2000, Integrating contextual information with per-pixel classification for improved land cover classification. *Remote Sensing of Environment*, **71**, 282–296.

Swain, P.H. and Davis, S.M., 1978, *Remote Sensing: The quantitative approach*, New York, Mc-Graw-Hill Book Company, 396 pp.

Turner, J., Kelly, J. and Newman, L.A., 1978, Soil nutrient-vegetation relationships in the Eden area, NSW. II. Vegetation-soil associations. *Australian Forestry*, **41**, 223–231.

Twery, M., Elmes, G.A. and Yuill, C.B., 1991, Scientific exploration with an intelligent GIS: Predicting species composition from topography. *AI Applications*, **5**, 45–53.

Varjo, J., 1997, Change detection and controlling forest information using multi-temporal Landsat TM imagery. *Acta Forestalia Fennica*, 0001-5636, 258, 64

Vogelman, J.E. and Rock, B.N., 1988, Assessing forest damage in high-elevation coniferous forests in Vermont and New Hampshire using thematic mapper data. *Remote Sensing of Environment*, **24**, 227–246.

Williams, D.L. and Nelson, R.F., 1986, Use of remotely sensed data for assessing forest stand conditions in the eastern United States. *Transactions on Geoscience and Remote Sensing*, **GE-24**, 130–138.

Wolter, P.T., Mladenoff, D.J., Host, G.E. and Crow, T.R., 1995, Improved forest classification in the northern Lake States using multi-temporal Landsat imagery. *Photogrammetric Engineering and Remote Sensing*, **61**, 1129–1143.

Woodcock, C.E. and Harward, V.J., 1992, Nested-hierarchical scene models and image segmentation. *International Journal of Remote Sensing*, **13**, 3167–3187.

Woodcock, C.E., Collins, J.B., Gopal, S., Jakabhazy, V.D., Li, X., Macomber, S., Ryherd, S., Wu, Y., Harward, V.J., Levitan, J. and Warbington, R., 1994, Mapping forest vegetation using Landsat TM imagery and a canopy reflectance model. *Remote Sensing of Environment*, **50**, 240–254.

Wulder, M., Nieman, K.O. and Goodenough, D.G., 2000, Local maximum filtering for the extraction of tree locations and basal area from high spatial resolution imagery. *Remote Sensing of Environment*, **73**, 103–114.

7

Application of remote sensing and geographic information systems in wildlife mapping and modelling

Jan de Leeuw, Wilbur K. Ottichilo, Albertus G. Toxopeus and
Herbert H.T. Prins

ABSTRACT

Wildlife management requires reliable and consistent information on the
abundance, distribution of species and their habitats as well as threats. This article
reviews the application of remote sensing and GIS techniques in wildlife
distribution and habitat mapping and modelling.

7.1 INTRODUCTION

The main purpose of wildlife conservation is to maintain maximum plant and
animal diversity through genetic traits, ecological functions and bio-geo-chemical
cycles, as well as maintaining aesthetic values (IUCN 1996). This has been
achieved to a certain extent through the creation of parks and reserves in different
parts of the world. These areas are set aside and managed to protect individual
plant and animal species, or more commonly of assemblages of species, of habitats
and groups of habitats. Different criteria are used in the establishment of parks and
nature reserves. Ideally they should comprise communities of plants and animals
that are in balance, and exhibit maximum diversity (Jewell 1989). However, some
areas have been designated as parks or reserves based on high-profile species only
or because they form a habitat for endangered or endemic plants or animals or are
unique natural landscapes. Many parks are declared for purposes other than wildlife
conservation.

For over a century national parks and reserves have been the dominant
method of wildlife conservation (Western and Gichohi 1993). Because most of
these areas are not complete ecological units or functional ecosystems in
themselves, they have experienced a range of management problems. The main
problem is the general decline in plant and animal diversity (Western and Gichohi
1993). A new approach is thus the 'ecosystem approach' to promote biological
diversity outside the traditional protected areas (Prins and Henne 1998).

Wildlife, and its conservation, is in crisis. Unprecedented and increasing loss
of native species and their habitats has been caused by different human activities.
Management strategies have focused mainly on single species and protected areas.

Immediate conservation is required particularly for areas outside the protected area system, which have rich wildlife resources. However, this action is hampered by lack of information and knowledge about species abundance, species distributions and factors influencing their distributions in these areas. Also there is general lack of understanding about the ecological, social and cultural processes that maintain diversity in different areas or ecosystems, i.e. of wildlife conservation at a landscape scale.

In this chapter, the application of remote sensing (RS) and geographic information system (GIS) in the collection and analysis of wildlife abundance and distribution data suitable for conservation planning and management are examined. Section 7.2 briefly examines issues related to wildlife conservation and reserve management. Section 7.3 reviews the techniques used in mapping wildlife distributions and their habitats. Resources required by wild animals to fulfil their life cycle needs are described in section 7.4. Section 7.5 reviews the application of GIS in mapping and modelling suitability for wildlife and factors influencing their distribution. Modelling of species-environment relationship is discussed in section 7.6. A future innovative potential of the use of RS and GIS in the collection, analysis and modelling of wildlife abundance and distribution is briefly discussed in section 7.7.

7.2   WILDLIFE CONSERVATION AND RESERVE MANAGEMENT

With the exponential growth of human populations, and the consequent demand on natural resources, the Earth is being transformed from large expanses of natural vegetation towards a patchwork of natural, modified and man-made ecosystems. Faced with this reduction, fragmentation or complete disappearance of their specific habitat, many wildlife species have suffered reductions in their numbers or range, or have become extinct. The underlying factors responsible may be classified as those with a direct negative effect, such as hunting, fishing, collection or poaching, and those indirectly detrimental to wildlife through impact on their habitat. Among these, the alteration and loss of habitat is considered the greatest threat to the richness of life on Earth (Meffe and Carroll 1994).

Over the past century, conservation efforts have concentrated on the acquisition and subsequent protection of critical wildlife habitat. Today, approximately 7.74 million km^2 or 5.19% of the world's land surface is designated and protected as parks or reserves (WCMC 1992). Many of these parks and reserves, however, were created as attractions with geological or aesthetic appeal rather than for biological conservation. In general, they are remnants of lands with marginal agricultural value, while highly productive lands tend to be underrepresented (Meffe and Carroll 1994). The International Union for Conservation of Nature (IUCN) recommended the preservation of a cross-section of all major ecosystems and called for protection of 13 million km^2 of the Earth's surface (Western 1989).

Once established, reserves do not necessarily guarantee the conservation of wildlife, because various processes operating within their boundaries might negatively affect wildlife. In many cases, protection within reserve remains marginal at best, exposing wildlife to incompatible land uses such as livestock grazing, mining, agriculture or logging. Some species are vulnerable to poaching or over exploitation. In addition, exotic diseases or invasive species may impact wildlife populations (Prins 1996). Modification of environmental conditions including the availability of resources such as water points for livestock, may change the balance amongst native species, advantaging some and disadvantaging others. Visitors may exert a negative impact on wildlife or their environment, particularly in highly frequented areas or where sensitive species occur.

Traditionally, wildlife management focussed on the maintenance of some desired state of the resource base within the reserve, while controlling factors negatively impacted on wildlife and the resource base on which they depend. Such internal management does however not guarantee sustainable wildlife conservation. Biological and physical processes in the surrounding areas may negatively impact on populations residing in the reserve (Janzen 1986; Prins 1987). Fragmentation of wildlife habitat outside reserves for instance is considered a potentially important factor negatively affecting wildlife within (Meffe and Caroll 1994). Wildlife populations in reserves might be too small to persist on their own and depend for their long-term survival on interbreeding with other sub-populations inhabiting similar habitat outside. Fragmentation of the habitat outside would increase the isolation of the population inside the reserve and increase the probability that it will go extinct (Soulé 1986).

Nowadays many reserves are confronted with increased intensity of land use at their periphery. Therefore, successful wildlife management requires the provision and maintenance of optimal conditions both within and outside reserve boundaries. Species with large territories may be at risk when individuals cross reserve boundaries, e.g. grizzly bears may be shot by rangers when posing a threat to cattle. Successful wildlife management requires appropriate data on wildlife especially data on spatial and temporal abundance and distribution. Remote sensing and GIS techniques are increasingly being used in the collection and analysis of these data as well as the monitoring and overall management of wildlife.

7.3 MAPPING WILDLIFE DISTRIBUTION

Geographic information on the distribution of wildlife populations forms a basic source of information in wildlife management. Most commonly, distribution is derived from observations in the field of the animal species or their artefacts. Radio-telemetry and satellite tracking have been used (Thouless and Dyer 1992) to record the distribution of a variety of animal species.

Aerial survey methods based on direct observation augmented by use of photography have been used to map the distribution of various taxonomic groups such as mammals (Norton-Griffiths 1978), birds (Drewien *et al.* 1996; Butler *et al.* 1995) and sea turtles and marine mammals (Wamukoya *et al.* 1995). Aerial photography has been used to map the distribution particularly of colonial species such as birds (Woodworth *et al.* 1997) or mussels (Nehls and Thiels 1993).

GIS is increasingly used for mapping wildlife density and distribution derived from ground or aerial survey observations (Butler *et al.* 1995; Said *et al.* 1997). For example, Figure 7.1 displays the distribution of wildebeest in the Mara ecosystem in Narok district (Said *et al*. 1997). McAllister *et al*. (1994) used GIS to analyze the global distribution of coral reef fishes on an equal-area grid.

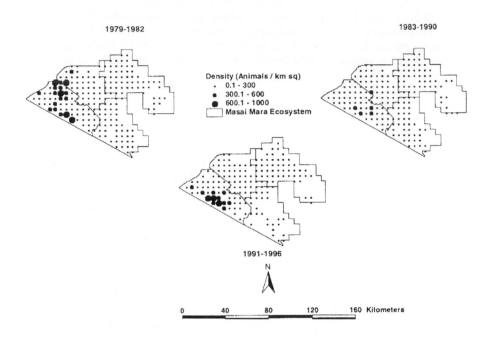

Figure 7.1: Spatial distribution and average density (N.km²) of wildebeest in the Masai Mara ecosystem, Narok District, Kenya for the period 1979–1982, 1983–1990 and 1991–1996. The density was calculated on 5 by 5 km sub-unit basis.

Satellite remote sensing undoubtedly has a potential for mapping of animal distribution, but successful applications seem to be few. Mumby *et al*. (1998a) mapped coral reefs using aerial photography and remote sensing imagery. For mapping of nine reef classes, they reported an overall accuracy of 37 per cent for Landsat TM and 67 and 81 per cent with aerial photography and an airborne CASI hyperspectral scanner respectively. Mumby *et al*. (1998b) reported that classification accuracy could be significantly increased by compensation for light attenuation in the water column and contextual editing. Thermal scanners have been used to determine the presence and/or numbers of animals not readily observable, such as beavers and muskrats in their lodges during winter (Intera Environmental Consultants 1976). They have also been used in Canada to count bison, moose, deer and elk in comparison with aerial and ground counts (Intera Environmental Consultants 1976). The main drawback is error emanating from hot spots such as solar heated objects, vacated sleeping spots and non-target animals.

A number of species such as termites, earth worms, or shellfish increase the roughness of the substrate, either through their exoskeleton or through their impact on soil micro-topography. Radar, being sensitive to such micro-relief (Weeks *et al.* 1996, Van Zyl *et al.* 1991), could potentially be applied to map such animal populations.

Hence, successful satellite-borne remote sensing applications seem to be restricted to cases where species modify their environment to such extent that their impact on the environment can be detected by a sensor. It is envisaged that the ability to map animal distribution in this way will be greatly enhanced by the advent of high spatial resolution remote sensing platforms.

7.4 MAPPING WILDLIFE RESOURCE REQUIREMENTS

Resources used by animals include those material goods required to fulfil their life cycle such as food, drinking water, nesting sites, shelter etc. Vegetation maps tend to be used to map the spatial distribution of these resources (with the exclusion of drinking water) (Flather *et al.* 1992). In some studies, the distribution of a species has been related directly to the classes or map units of these vegetation maps (August 1983). Here it remains undetermined whether the animal is located in one vegetation class or another because of the availability of food resources, shelter, nesting or a combination of those. Researchers and managers have converted the information provided by a vegetation map into the spatial distribution of the individual resources. Pereira and Itami (1991) used prior knowledge on the feeding ecology of the Mt Graham squirrel and seed productivity for various conifer species, to derive a food productivity map from a land cover map containing information on dominant tree species.

Articles presenting vegetation maps[1] or describing the techniques to produce them frequently stress the utility of such maps for wildlife or faunal management. Typically, vegetation maps contain thematic information on physiognomy, species composition or some other vegetation attributes (see for example Loth and Prins 1986). A survey on the thematic content of a sample of 169 rangeland vegetation maps, mostly from the African continent (Waweru 1998), revealed that 115 (68 per cent) and 69 (40 per cent) maps included information on vegetation physiognomy and species composition respectively. Forty out of the 169 maps (24 per cent) provided information on vegetation biomass while only two maps (1.2 per cent) provided explicit information on vegetation quality.

Although they are the most frequently mapped attributes, one might question whether vegetation physiognomy and species composition would be the most appropriate ones from a wildlife management perspective. Wildlife managers might well prefer information on the quantity and quality of food resources, which are considered major factors determining the distribution of animals.

Remote sensing has been applied to quantify the spatial distribution of vegetation biomass (Box *et al.* 1989; Prince 1991; Hame *et al.* 1997). This quantification is mainly done by means of Normalized Difference Vegetation Index

---

[1] For techniques for preparation of vegetation maps the reader is referred to Chapter 6. This section focuses on the application of vegetation maps to wildlife management.

(NDVI), or 'greenness index' (Tucker 1979) (see Chapter 4 for details). Annually integrated NDVI was shown by Goward *et al.* (1985) to be related to biome averages of annual net primary production (NPP). Prince (1991) demonstrated that there is a strong linear relationship between the satellite observation of vegetation indices and the seasonal primary production. Wylie *et al.* (1991) determined the relationship between time-integrated normalized difference vegetation index statistics and total herbaceous biomass through regression analysis. He concluded that availability of several years of data makes it possible to identify the temporal and spatial dynamics of vegetation patterns within the Sahel of Niger in response to year to year climatic variations. Although the NDVI appears to be a useful index of some surface phenomena, it is still not certain what biological phenomena the NDVI actually represents (Box *et al.* 1989). NDVI values based on the current NDVI products are not reliable in complex terrain (high mountains, coastal areas, irrigated areas in dry climates, etc.) due to mixed pixels. The NDVI values do not fall to zero in deserts or over snow cover, due to background effects (Box *et al.* 1989). However, current NDVI data seem reliable elsewhere, at least for annually integrated totals (Prince and Tucker 1986).

Many studies have been undertaken to relate NDVI to crop production (e.g. Groten and Ilboudo 1996) or grass biomass production (e.g. Prince and Tucker 1986). However, there are very few studies that have attempted to relate NDVI to animal distributions (e.g. Muchoki 1995; Omullo 1996; Oindo 1998).

Drinking water constitutes a critical resource to wildlife, particularly in arid and semi-arid zones. Hence, one would expect water dependent animals to be close to watering points. In studies in the Tsavo and Mara ecosystem of Kenya, Omullo (1995), Rodriguez (1997) and Oindo (1998) all reported significant relationships between the distribution of various wildlife species and the distance to permanent water points.

7.5 MAPPING AND MODELLING HABITAT SUITABILITY FOR WILDLIFE

In this section, habitats and habitat maps are described first. This is followed by a discussion about mapping of habitat suitability for wildlife, accuracy of the suitability maps and factors influencing wildlife distributions.

7.5.1 Habitats and habitat maps

Information and maps on wildlife distributions are essential for wildlife management. In many cases however, management interventions focus on the resource base on which the animals depend, rather than on the animals themselves as the vegetation or habitat is managed more easily than the animals themselves. Wildlife management organizations therefore traditionally displayed a strong interest in the mapping of resources relevant to wildlife. The underlying idea was that maps displaying the resource base could assist to identify areas suitable for wildlife.

Vegetation maps as well as so-called habitat maps have been used for this purpose. Traditionally, the term habitat has been defined either as the place or area where a species lives and/or as the (type of) environment where a species lives, either actually or potentially (Corsi *et al.* 2000). In all of the definitions reviewed by Corsi *et al.* (2000), the term habitat has been defined as the property of a specific species. Consequently, it can only be used in association with a name of a species, e.g. flamingo or tsetse habitat. This corresponds to the original use of the word, which was derived from *habitare* (to inhabit) in old Latin descriptions of a species. Hence, one would expect a habitat map to display information on the distribution of the habitat of a specific species. This, however, is not the case; habitat maps display information on the distribution of vegetation types or land units. For some intractable reason, these map units have been called habitats, e.g. a riverine or a woodland habitat, which is clearly a wrong but well-established terminology. In conclusion, habitat maps do not pertain to a specific species but refer to vegetation types or land units.

Use of the term habitat is not restricted to habitat maps. It has proliferated into the literature dealing with the assessment of suitability of land for wildlife. In habitat evaluation, habitat suitability index models and habitat suitability maps the term refers to units of land rather than to specific species.

The various meanings of the term habitat lead to ambiguity, for instance when used in the context of suitability assessment. According to the definition, above all habitat would by definition be suitable and unsuitable habitat would be a contradiction in terms. Areas unsuitable for a species would therefore have to be considered as non-habitat. When used in the second meaning, however, all land would be labeled as habitat, irrespective whether it would be suitable for a species or not. In this chapter, the term habitat is avoided whenever possible, and when applied it is used in relation to a specific wildlife species. The more neutral terms 'wildlife suitability model' and 'wildlife suitability map' are adopted.

7.5.2 Mapping suitability for wildlife

A wildlife suitability map is defined as a map displaying the suitability of land (or water) as a habitat for a specific wildlife species. Since the early 1980s, remote sensing has been used to localize the distribution of areas suitable for wildlife. Cannon *et al.* (1982), for instance, used Landsat MSS to map areas suitable for lesser prairie chicken. Wiersema (1983) mapped snow cover using Landsat MSS to identify snow free south facing slopes forming the winter habitat of the alpine ibex. Hodgson *et al.* (1987) used Landsat TM for mapping wetland suitable for wood stork foraging. More recently, Congalton *et al.* (1993) used a Landsat TM based vegetation map to classify the suitability of land for deer. Rappole *et al.* (1994) used Landsat TM to assess habitat availability for the wood thrush.

These studies depended on a vegetation map, derived from remote sensing, as the only explanatory variable. The assumption was that mapping units efficiently reflect the availability of resources and other relevant environmental factors determining suitability. However, the suitability of land for wildlife may be determined by more than one factor. A single explanatory variable, such as a vegetation map or a land-unit map, does not effectively represent such multiple factors, especially when they were poorly correlated with each other. This is frequently the case; the distribution of good quality grazing areas in arid zones, for instance, does not necessarily correspond to the availability of drinking water resources (Toxopeus 1998). In such cases, where factors are unrelated, GIS will be useful, since separate data layers may be combined in order to provide information on the distribution of independent landscape attributes.

In the second half of the 1980s, wildlife suitability maps integrating various explanatory variables were implemented in a GIS environment. Figure 7.2 shows a scheme of suitability mapping in a GIS context (see also Chapter 2 for a definition of model terms). Such a scheme consists of a suitability model that allows one to predict the suitability of land for a specific species, given a number of landscape attributes. Additionally, it contains a number of spatial databases describing the distribution of these landscape attributes. The suitability model is then used to process these spatial databases to generate a suitability map (Toxopeus 1996).

GIS-based habitat studies generally combine information on vegetation type or some other land cover descriptor, with other land attributes reflecting the resource base as well as other relevant factors. A model for Florida scrub jay developed by Breiniger *et al.* (1991), for instance, included vegetation type and soil drainage to discriminate primary habitat, secondary habitat and unsuitable areas. A more detailed model for the same species (Duncan *et al.* 1995) included seven attributes, all related to land cover.

Herr and Queen (1993) developed a GIS-based model to identify potential nesting habitat for cranes in Minnesota. A significant relation was observed to cover type, and two disturbance-related factors: distance to roads and distance to houses. Clark *et al.* (1993) included seven land attributes: land cover, elevation, slope, aspect, distance to roads, distance to streams and forest cover diversity to predict habitat suitability for black bear.

7.5.3 Accuracy of suitability maps

Wildlife suitability maps and their underlying suitability models have been criticized because of their assumed poor accuracy (Norton and Williams 1992). The maps produced by these models have rarely been validated (Stoms *et al.* 1992; Williams 1988), although this had clearly been advised in the habitat evaluation procedures (USFWS 1981). The accuracy of a wildlife suitability map depends on how well the output corresponds to reality (Figure 7.2). This accuracy is determined by two different sources of error. The first source of error is the spatial database, which comprise both geometric and thematic errors. The second source of error is the habitat suitability model. The accuracy of suitability models depends on the selection of the relevant variables and an unbiased estimation of the model parameters.

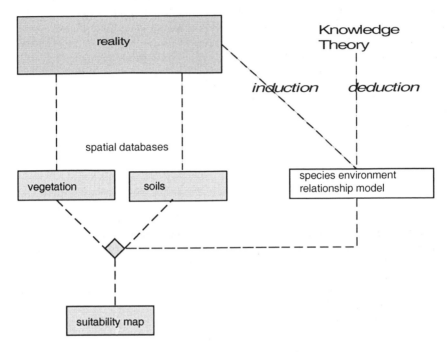

Figure 7.2: Scheme for GIS based suitability mapping.

Accuracy assessment of wildlife suitability models has been discussed in Morrison *et al.* (1992), while Corsi *et al.* (2000) provides a review of potential techniques to assess the accuracy of wildlife suitability maps. Skidmore (1999), Janssen and Van der Wel (1994) and Congalton (1991) give general discussions on techniques to assess map accuracy. These map accuracy assessment techniques require separate data sets for validation of the model developed. Verbyla and Litvaitis (1989) indicated that, in wildlife suitability studies, the number of samples may be too small and described resampling methods to overcome this problem.

In accuracy assessment, the predicted suitability is tabulated against observations on presence and absence of the animal species. Morrison *et al.* (1992) reviewed the reasons why animals would not be recorded in suitable areas (Type 1 error) or would be observed in areas considered unsuitable (Type 2 error). Most animal species are mobile, hence suitable land may not be temporarily occupied, while animals may pass through lands otherwise unsuitable to them. Furthermore, animals may be locally extinct. Animals differ in this respect from plant species or land cover and, because of this, accuracy matrices for wildlife-suitability-maps may yield relatively low accuracy values. We argue that such low accuracy values do not necessarily imply poor model performance. After all, the model predicts suitability rather than presence or absence. Besides, models with a low accuracy may still contain ecologically relevant information.

The potential of a vegetation map to explain the distribution of wildlife depends on its map accuracy. The accuracy of the map information depends on the level of thematic detail. Anderson (1976) distinguished three different levels in land cover maps: Anderson level I corresponds to broad land cover classes such as forest versus grassland; Anderson level II gives a further separation according to broad species groups such as broad-leafed versus pine forest; Anderson level III includes detail such as vegetation types defined by species composition. Accuracy obtained for Anderson level I and II vegetation maps tend to be above 80 per cent, while Anderson level III maps remain below this accuracy level.

7.5.4 Factors influencing wildlife distribution

The actual distribution of animal species may be determined by a variety of environmental factors (Morrison *et al.* 1992). We categorize these into three broad classes; those describing the resource base, physico-chemical factors and factors related to human activities (Figure 7.3). Physico-chemical and anthropogenic factors may influence the distribution of wildlife either directly or indirectly through their impact on the resource base.

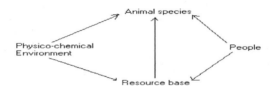

Figure 7.3: Scheme displaying the impact on the distribution of an animal species of three broad categories of environmental factors. People and the physical-chemical environment may exert a direct as well as an indirect impact through their influence on the resource base.

Johnson (1980) argued that selection of habitat by an animal species may occur at different spatial scales and proposed the following hierarchical order in the selection of habitat by an animal. First order selection corresponds to the geographic range of a species, second order selection to the home range of an animal or a social group, while third order selection pertains to utilization of resources within that home range.

It has been suggested by Diamond (1988) that different biophysical factors affect species richness at different scales. At the regional level, productivity and climatic zones determine species richness. This has been amply demonstrated in, for example, Rosenzweig (1995) but also by Veenendaal and Swaine (1998) in their analysis of the natural limits of the distribution of tree species from the West African rainforest. At the landscape level (or gamma level), productivity, climate (precipitation, temperature, growing season) play a role; this has been demonstrated for grazing herbivores in Africa (Prins and Olff 1998), but also for Gobi Desert rodents and even North Atlantic megafauna (fish, echinoderms and crustaceans) (Rosenzweig 1995). Even seasonality and plant phenological processes may play a role, for example, for primate assemblages in West Africa (Tutin and White 1998; see also Newbery *et al.* 1998). At the community level, the aforementioned factors play a role still, because the species assemblage at that level is a sample of the regional species pool. However, not all species of that pool will be found at the community level, often because of competition between species, and the smaller the area under scrutiny, the lower the number of species (Prins and Olff 1998). Lastly, at point or microhabitat level, the most important factors are soil moisture and soil nutrients, and, especially for plants, the light regime (Zagt and Werger 1998; Loth 1999). Especially at this level, chance effects, however, may dominate.

People and their associated activities may exert positive or negative influences on the distribution of wildlife. In the case of a negative impact, it may prevent the animals from occupation of otherwise suitable habitat. The potential number of human-induced disturbance factors is large and it would go beyond the scope of this chapter to list them all. However, most human-related disturbance factors do have one thing in common: their intensity or frequency diminishes with the distance from a human settlement or infrastructures used by people. Not surprisingly, therefore, distance has been used as an explanatory variable in many GIS-based wildlife distribution models. For instance, the areas mapped by Herr and Queen (1993) as suitable habitat for cranes were largely determined by distance to roads, buildings and agricultural lands. However, distance as such does not influence the distribution of the animals. Instead, an unknown variable (for instance, human disturbance) associated with distance would be the ultimate factor affecting the observed animal distribution (Prins and Ydenberg 1985). Distances should therefore be carefully interpreted and considered as factors reflecting associated human impact.

7.6   MODELLING SPECIES-ENVIRONMENT RELATIONSHIPS

The ability to model spatial distribution and change in distribution of wildlife is of considerable importance in wildlife management. Once spatial distribution can be adequately modelled, distribution and abundance may be monitored effectively over time. GIS can be effective in modelling animal distribution if the necessary data are available. However, data availability is currently the limiting factor in many areas.

Production of a suitability map requires a model to predict the suitability of land for a wildlife species given both a set of land attributes and also distribution of potential competitors. According to the source of knowledge on which they are based, such models may be classified as theoretical-deductive and empirical-inductive methods, based on the definitions in Chapter 2 (Figure 2.2). The former use theoretical considerations and existing knowledge to design a model, whereas the latter depend on knowledge on species environment relationships obtained through empirical research (Chapter 2).

Habitat suitability index (HSI) models, described by Atkinson (1985) as hypotheses about species-environment relationships based on the literature and opinions of experts, are an example of theoretical-deductive wildlife-environment relationship models. Hundreds of such models have been developed since the early 1980s (Atkinson 1985; Williams 1988) and several have been used to implement wildlife suitability maps in a GIS environment (Donovan *et al.* 1987; Duncan *et al.* 1995). Other deductive models have been presented by, for instance, Herr and Queen (1993) and Breininger *et al.* (1991) – see also Chapter 2. Deductive modelling, however, has severe drawbacks in wildlife ecology. For many species, knowledge about habitat requirements simply does not exist. However, expertise with respect to wildlife habitat requirements may be limited, biased or not be available (Kangas *et al.* 1993; Crance 1987).

Inductive modelling has been suggested to overcome these problems (Walker 1990; Walker and Moore 1988; Chapter 2). Inductive modelling is based on the analysis of data resulting in the generation of new knowledge and the formulation of new models. Here modelling goes from the specific case (field data) towards a generalization.

A variety of analytical techniques has been used to investigate species-environment relationships. These include logistic regression (Pereira and Itami 1991; Buckland and Elston 1993; Osborne and Tigar 1992; Walker 1990; Rodriguez 1997), discriminant analysis (Haworth and Thompson 1990), classification and regression trees (Walker and Moore 1988, Skidmore *et al.* 1996), canonical correlation analysis (Andries *et al.* 1994), supervised non-parametric classifiers (Skidmore 1998; Skidmore *et al.* 1996) and neural networks (Skidmore *et al.* 1997).

The distribution of a species may be related to many independent variables using a GIS. Initially this appears to be a panacea. However, one may become overwhelmed by the multitude of data layers available in a GIS. Many layers may be irrelevant to the problem at stake. The number of the independent variables included in the analysis could be reduced using *a priori* knowledge about the ecology of the species. Even then, however, many variables might be retained and frequently they will tend to be highly correlated. Such high mutual correlation is, for instance, a common phenomenon when using the various bands of a remote sensing image, and especially hyperspectral remote sensing (Skidmore and Kloosterman 1999; Van der Meer 1995) in wildlife suitability studies. Such collinearity may result in models that have a poor predictive power when extrapolated to non-surveyed sites. Osborne and Togar (1992) and Buckland and Elston (1993) used principal components analysis (PCA) and subsequently regressed the dependent variable against the principal components. Duchateau *et al.* (1997) used PCA and varimax rotation to reduce the dimensionality of the data and to identify a reduced set of climatic predictor variables. These were then regressed against the independent variable, the presence of outbreaks of a tick borne livestock disease. The reduction of the dimensionality was based on claims of superior performance when applied to an independent data set over models including a larger set of predictor variables. No attempt, however, has been made to verify this claim.

7.6.1 Static versus dynamic models

So far, wildlife suitability mapping techniques have been described in terms of static models, both for animal populations as well as for the environment (Table 7.1). In reality, both animal populations and resource bases tend to display highly dynamic behavior.

Table 7.1: Classification of GIS based models for wildlife management depend on whether a static or dynamic model has been used to map the resource base as well as whether the response of the animal population would be based on a static or dynamic model.

Resource base	Animal population	
	Static	Dynamic
Static	A	B
Dynamic	C	D

Breininger *et al.* (1998) studied the relationship between demographic characteristics to a habitat suitability index (HSI) map for Florida scrub-jay. Yearling production, breeder survival, demographic performance and jay density were significantly correlated to HSI. Pereira and Itami (1991) linked a static species-environment model to the current and an alternative state of the environment to assess the impact of the development of an astronomical observatory on the habitat of the Mt Graham squirrel. Such mapping of the suitability for wildlife does not capture the change over time due to succession, natural disturbances such as fire or storms, or human activities.

Prediction of the impact of human activities is relatively simple in case of such localized infrastructure projects. In many cases, it would be much more difficult to predict the location of future human impacts. Toxopeus (1995) predicted deforestation in Cibodas, Indonesia, using distance from settlements and accessibility of the terrain as predictor variables (Figure 7.4).

Kruse and Porter (1994) linked a dynamic resource base model to a habitat suitability model to evaluate the change in suitability of forest for wildlife over time under different management options.

A number of models have been published predicting the population dynamics of a wildlife species in response to a dynamic resource base. Two different types of models can be discerned, namely, non-spatially explicit models, and spatially explicit ones. For the spatially non-explicit models, we refer to those published on the reaction of wildebeest (*Connochaetus taurinus*) in the Serengeti in Tanzania to fluctuations in rainfall. Especially, variations in dry-season rainfall are important to understand population dynamics as rainfall determines the length of the growing season for the vegetation, which, in turn, determines animal condition and, thus, natality and mortality (Hilborn and Sinclair 1979; Mduma *et al.* 1998). Also the population dynamics of semi-wild Soay sheep on the Hebridian islands of Rhum and Hirta has successfully been modelled by Illius and Gordon (1999); fluctuations in rainfall and inclement weather determine the fluctuations and animal condition is linked to energy gains and, especially, losses.

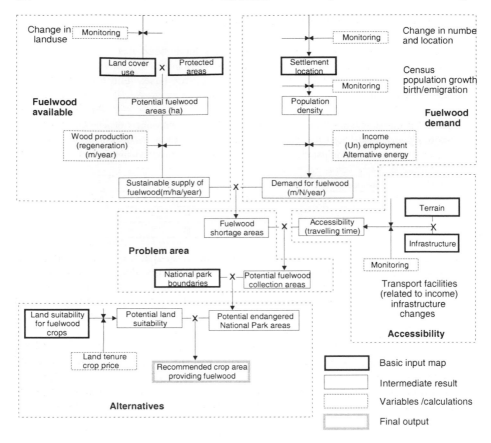

Figure 7.4: Scheme of a GIS model, applied to predict the fuelwood collecting areas in the Cibodas Biosphere reserve, West Java.

Dynamic wildlife-population models have also been linked to alternative states of the resource base. DeAngelis *et al.* (1998), for instance, predicted the reproductive performance of deer in the Everglades (in Florida, USA), based on the availability of resources as determined by the hydrological conditions under the current situation, and in an alternative scenario. Spatially explicit models linking the dynamics of the resource base to population dynamics are still rare. Central to these models are the feedback mechanisms between animal consumption, plant production and competition between plant species (Van Oene *et al.* 1999). Indeed, competition between plant species determines vegetation composition, and this, in turn, determines suitability for herbivores. Population dynamics of three large grazers (Red deer *Cervus elaphus*, Heck cattle, and Konik horses) and vegetation composition has thus been spatially modelled for the wetland 'Oostvaardersplassen' in the Netherlands (Groot Bruinderink *et al.* 1999).

7.6.2 Transferability of species – environment models

For a limited number of species, models have been developed for a particular site. However, the transferability of such site-level models to other areas remains unknown, and may lead to biased results. Consequently, application of a site-level model to a whole region is not recommended (Risser *et al.* 1984). So far, only a few papers have addressed the problem of transferability of wildlife suitability models, in geographically restricted areas. Thomas and Bovee (1993) investigated transferability of habitat suitability criteria between two rivers in Colorado. Homer *et al.* (1993) concluded that a model developed in a sub-area of a 2740 km^2 county provided reliable predictions when transferred to two other sub-areas in the same county.

Once established, a wildlife manager will be tempted to apply a suitability model and derived maps to the future. This assumes that the underlying species-environment model would remain unchanged through time. However, this will not be the case when relevant ecological factors affecting animal distribution have not been included in the model. Rainfall in arid and semi-arid zones, for example, is known to vary in both time and space. It is noteworthy that none of the cited articles related animal distribution to spatial pattern of rainfall prior to or during the study. Instead, animal distribution tends to be related to long-term averages of climatic variables. Oindo (1998) used NOAA-AVHRR NDVI instead of rainfall data to reflect spatial variation in vegetation phenology. He related NDVI, together with other landscape attributes, to the observed distribution of Topi around Masai Mara Reserve, Kenya. NDVI, however, did not significantly explain the observed distribution. Oindo (1998) reported that a suitability model for Topi, developed for a particular year, correctly predicted the distribution of the species in other years, indicating that the model may be transferred over time.

7.7 INNOVATIVE MAPPING OF WILDLIFE AND ITS PHYSICAL ENVIRONMENT

Various aspects of the physical environment have been used in wildlife suitability studies. It is beyond the scope of this chapter to review all of these and discuss methods to map them. Here we highlight a selection of factors for which innovative methods for mapping wildlife distribution appeared in the past few years.

Climatic databases, derived through interpolation from point-based observations, have been used to map distribution at continental scales (Prins and Olff 1998). At larger scales, micro-topographical climatic and soil variation becomes more prominent (Varekamp *et al.* 1996). Slope and aspect, which determine the local moisture regime through their impact on the solar radiation balance, have frequently been in suitability models. Nowadays, GIS-based models for mapping solar radiation in relation to topography are available (Kumar *et al.* 1997), but so far these have apparently not been used for wildlife suitability studies.

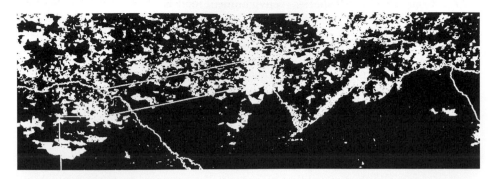

Figure 7.5: Map indicating bush fires (burn scars) in August, 1996, based on NOAA-AVHRR data, of the Caprivi region in Namibia (source: Mendelsohn and Roberts, 1997).

Fire occurs in many ecosystems (Huston 1979), and affects wildlife populations. Recent forest fires for instance exerted a negative impact on orangutan populations in Kalimantan, Indonesia. Thermal remote sensing allowed mapping and monitoring the distribution of fire. Historic fire-maps based on NOAA-AVHRR thermal bands are available since the early 1990s (Figure 7.5).

Flooding may influence the distribution of wildlife habitat through its impact on the resource base animals depend on. Radar imagery has successfully been applied to map flooded areas (Richards *et al.* 1987; Pope *et al.* 1992; Imhof 1986), even underneath closed vegetation canopies. DeAngelis *et al.* (1998) used information on flooding derived from radar to predict the reproductive rate of deer in the Everglades.

A critical gap remains between satellite data and the many varieties of field observations (Miller 1994). Currently verification of broad-scale mapping efforts using field survey data is still a problem in our attempts to map natural and human induced features. A new approach that combines GPS and videography offers a practical method to validate and classify TM imagery to produce vegetation maps (Graham 1993).

Species distribution mapping is an increasingly important part of ecological science (Miller 1994). The equal-area grid arrangement is a useful framework for representing species presence/absence data and for analyzing species distribution patterns. This approach is suitable mainly for developing countries where the primary sources of data on species spatial distribution are very often presence/absence data. This approach is demonstrated in several recent studies of birds in East Africa (Pomeroy 1993).

7.8 CONCLUSIONS

The potential for the use of RS and GIS technologies in wildlife mapping, natural resource planning and management are large. These technologies are currently fully developed and they are increasingly being applied in natural resource mapping, planning and management. However, their application, particularly in developing countries, is still limited by lack of appropriate scale of data, hardware, software and expertise.

Future research in wildlife modelling should focus on developing more realistic dynamic models of wildlife in space and time. Since the ecosystems to be modelled can be significantly affected by stochastic events and the responses of wildlife are non-linear in form, models must be dynamic and aim to provide predictions of known precision that are testable.

Given that the two most common questions asked by wildlife managers are likely to remain 'where and in what abundance does it occur?' and 'what will happen to it if....?', continued research should be directed to refining models that can best answer these questions (Norton and Possingham 1993).

7.9 REFERENCES

Anderson, J.R. 1976, *A land use and landcover classification system for use with remote sensor data*. Washington, D.C: United States Government Printing Office, Version 2.

Andries, A.M., Gulinck, H. and Herremans, M., 1994, Spatial modelling of the barn owl Tyto alba habitat using landscape characteristics derived from SPOT data. *Ecography*, **17**, 278–287.

Atkinson, S.F., 1985, Habitat-based methods for biological impact assessment. *The Environmental Professional*, **7**, 265–282.

August, P.V., 1983, The role of habitat complexity and heterogeneity in structuring tropical mammal communities. *Ecology*, **64**, 1495–1507.

Breininger, D.R., Provancha, M.J. and Smith, R.B., 1991, Mapping Florida Scrub Jay habitat for purposes of land-use management. *Photogrammetric Engineering and Remote Sensing*, **57**, 1467–1474.

Breininger, D.R., Larson, V.L., Duncan, B.W. and Smith, R.B., 1998. Linking habitat suitability to demographic success in Florida scrub-jays. *Wildlife Society Bulletin*, **26**, 118–128.

Box, E.O., Holben. B.N. and Kalb, V., 1989, Accuracy of the AVHRR vegetation Index as a preditor og biomass, primary productivity and net CO_2 flux. *Vegetatio*, **80**, 71–89.

Buckland, S.T. and Elston, D.A., 1993, Empirical models for the spatial distribution of wildlife. *Journal of Applied Ecology*, **30**, 478–495.

Butler, R.W. and Kaiser, G.W., 1995, Migration chronology, sex ratio, and body mas of least sandpipers in British Columbia. *Wilson Bull.*, **107**, 413–422.

Cannon, R.W., Knopf, F.L. and Pettinger, L.R., 1982, Use of Landsat data to evaluate lesser prairie chicken habitats in western Oklahoma. *Journal of Wildlife Management*, **46**, 915–922.

Clark, J.D., Dunn, J.E. and Smith, K.G., 1993, A multivariate model of female black bear habitat use for a geographic information system. *Journal of Wildlife Management*, **57**, 519–526.

Congalton, R., 1991, A review of assessing the accuracy of classifications of remotely sensed data. *Remote Sensing of Environment*, **37**, 35– 46.

Congalton, R.G., Stenback, J.M. and Barrett, R.H., 1993, Mapping deer habitat suitability using remote sensing and geographic information systems. *Geocarto International*, **3**, 23–33.

Corsi, F., De Leeuw, J. and Skidmore, A.K., 2000, Modelling species distribution with GIS. In: Boitani, L. and Fuller, T.K. (eds.), *Research Techniques in Animal Ecology Controversies and Consequences*. New York, Colombia University Press, , 389–434.

Crance, J.H., 1987, *Guidelines for using the Delphi Technique to develop habitat suitability index curves*. Biological report 82, National Ecology Center, Washington DC, 22

De Angelis, D.L., Gross, L.J., Huston, M.A., Wolff, W.F., Fleming, D.M., Comiskey, E.J. and Sylvester, S.M., 1998, Landscape modeling for Everglades ecosystem restoration. *Ecosystems*, **1**, 64–75.

Diamond, J., 1988, Factors controlling species diversity: Overview and synthesis. *Annals of the Missouri Bortanical Garden*, **75**, 117–129.

Donovan, M.L., Rabe, D.L. and Olson, C.E., 1987, Use of Geographic Information Systems to develop habitat suitability models. *Wildlife Society Bulletin*, **15**, 574–579.

Drewien, R.C., Brown, W.M. and Benning, D.S., 1996, Distribution and abundance of sandhill cranes in mexico. *Journal of Wildlife management*, **60**, 270–285.

Duchateau, L., Kruska, R.l. and Perry, B.D., 1997, Reducing a spatial database to its effective dimensionality for logistic-regression analysis of incidence of livestock disease. *Preventive veterinary medicine*, **33**, 205–216.

Duncan, B.W., Breininger, D.R., Schmalzer, P.A. and larson, V.L., 1995, Validating a Florida Scrub Jay Habitat Suitability Model, using Demography Data on Kennedy Space Center. *Photogrammetric Engineering and Remote Sensing*, **61**, 1361–1370.

Flather, C.H., Bredy, S.J. and Inkley, D.B., 1992, Regional habitat appraissals of wildlife communities: a landcscape-level evaluation of a resource planning model using avian distribution data. *Landscape Ecology*, **7**, 137–147.

Graham, L., 1993, Airborne video for near real-time natural resource applications. *Journal of Forestry*, **91**, 28–32.

Green, R.O. 1992, *Summaries of the Third Annual JPL Airborne Geoscience Workshop*, JPL 92–14, Vol. 1. Jet propulsion Lab., California Institute of Technology, Pasadena, California.

Groot Bruinderink, G.W.T.A., Baveco, J.M., Kramer, K., Kuiters, A.T., Lammertsma, D.R., Wijdeven, S., Cornelissen, P., Vulink, J.T., Prins, H.H.T., van Wieren, S.E. de Roder F. and Wigbels, V., 1999, *Dynamische interacties tussen hoefdieren en vegetatie in de Oostvaardersplassen*. IBN-rapport 436. Alterra/Wageningen University and Research Centre, Wageningen.

Groten, S.M.E., and Ilboudo, J., 1996, *Food security monitoring, Burkina Faso*. BCRS, Netherlands Remote Sensing Board (NRSP rapport; 95–32).

Goward, S.N., Tucker, C.J., and Dye, D.G., 1985, North American vegetation patterns observed with the NOAA advanced very high resolution radiometer. *Vegetatio*, **64**, 3–14.

Hame, A., Salli, K. Anderson, K. and Lohi, A., 1997, *International Journal of Remote Sensing*, **18**, 3211–3243.

Haworth, P.F. and Thompson, D.B.A., 1990, Factors associated with the breeding distribution of upland birds in the south Pennines, England. *Journal of Applied Ecology*, **27**, 562–577.

Herr, A.M. and Queen, L.P., 1993, Crane habitat evaluation using GIS and remote sensing. *Photogrammetric Engineering and Remote Sensing*, **59**, 1531–1538.

Hilborn, R. and Sinclair, A.R.E., 1979, A simulation of the wildebeest population, other ungulates and their predators. In: A.R.E. Sinclair and M. Norton-Griffiths (eds.). *Dynamics of an ecosystem*. Chichago, University of Chicago Press. pp. 287–309

Hodgson, M.E., Jensen, J.R., Mackey, H.E. Jr., and Coulter, M.C., 1987, Remote Sensing of wetland habitat: A wood stork example. *Photogrammetric Engineering and Remote Sensing*, **53**, 1075–1080.

Homer, C.G., Edwards, T.C. JR., Ramsey, R.D. and Prince, K.P., 1993, Use of remote sensing methods in modelling sage grouse winter habitat. *Journal of Wildlife Management*, **57**, 78–84.

Huston, M. 1979, A general hypothesis of special diversity. *The American Naturalist*, **113**, 81–101.

Illius, A.W. and Gordon, I.J., 1999, Scaling up from functional response to numerical response in vertebrate herbivores. In: Olff, H. Brown V.K. and Drent R.H. (ed.) *Herbivores: between plants and predators*. BES. Symp. Vol. 38, 397– 426. Oxford, Blackwells Scientific.

ImHoff, M.L., 1986, Space-borne radar for monsoon and storm induced flood control planning in Bangladesh: a result of the shuttle imaging radar-B program. *The Science of Total Environment*, **56**, 277–286.

Intera Environmental Consultants Ltd., 1976, *Report on thermal sensing of ungulates in Elk Island National Park*, Calgary, Canada. In ILCA Monograph 4 (1981), Addis Ababa, Ethiopia.

IUCN, 1996, *1996 IUCN red list of threatened animals*. IUCN, Gland, Switzerland.

Janssen, L.L.V. and Van der Wel, F.J.M., 1994, Accuracy Assessment of Satellite Derived land-Cover Data: A Review. *Photogrammetric Engineering and Remote Sensing*, **60**, 419–424.

Janzen, D.H., 1986, The eternal external threat. In: Soule, M.E. (ed.) *Conservation biology, the science of scarcity and diversity*. Sunderland Sinauer. pp. 286–303.

Jewell, N., 1989, An evaluation of multi-SPOT data for agriculture and land use mapping in the United Kingdom. *International Journal of Remote Sensing*, **10**, 939–951.

Johnson, D.H., 1980, The comparison of usage and availability measurements for evaluating resource preference. *Ecology*, **61**, 65–71.

Kangas, K., Karsikko, J., Laasonen, L. and Pukkala, T., 1993, A method for estimating the suitability function of wildlife habitat for forest planning on the basis of expertise. *Silva Fennica*, **27**, 259–268.

Kruse R.L. and Porter, W.F., 1994, Modelling changes in habitat conditions in northern hardwoods forests of the Aircondack Mountains in New York. *Forest Ecology and Management*, **70**, 99–112.

Kumar, L., Skidmore, A.K. and Knowles, E., 1997, Modelling topographic variation in solar radiation in a GIS environment. *International Journal of Geographical Information Science*, **11**, 475–497.

Loth, P.E., 1999, *The vegetation of Manyara: scale-dependent states and transitions in the African Rift Valley*. PhD thesis, Wageningen University.

Loth, P.E. and Prins H.H.T., 1986, Spatial patterns of the landscape and vegetation of Lake Manyara National Park, Tanzania. (Including a vegetation map 1 : 50,000). *ITC Journal* **1986 (2)**, 115 – 130.

McAllister, D.E., Schueler, F.W., Roberts, C.M. and Hawkins, J.P., 1994, *Mapping the Diversity of Nature*. In: Ronald I. Miller (ed.). London, Chapman and Hall.

Mduma, S., Hilborn, R. and Sinclair, A.R.E., 1998, Limits to exploitation of Serengeti wildebeest and implications for its management. In: D.M. Newbery, H.H.T. Prins and N.D. Brown (ed.) *Dynamics of tropical communities*. BES Symp. Vol. 37, 243 – 265. Oxford, Blackwells Scientific.

Meffe, G.K. and Carrol, C.R., 1994, *Principles of Conservation Biology*. In: Meffe, G.K., Carroll, C.R.and contributors (ed.). Sunderland, MA, Sinauer Associates.

Mendelsohn, J. and Roberts, C., 1997, *An Environmental Profile and Atlas of Caprivi*. Windhoek, Namibia, Directorate of Environmental Affairs.

Michelmore, F., 1994, *Keeping elephants on the map: Case studies of the application of GIS for conservation*. London, Chapman and Hall.

Miller, R.I. 1994, *Possibilities for the future. Mapping the Diversity of Nature*. London, Chapman and Hall.

Morrison, M.L., Marcot, B.G. and Mannan, R.W., 1992, *Wildlife-habitat relationships: concepts and applications*. Wisconsin, University of Wisconsin press, Madison.

Muchoki, C.H.K., 1995, *Progressive development of vegetation resources and its relationship to animal use in Amboseli ecosystem, Kenya: A remote sensing approach*. PhD Thesis, University of Aberdeen, Great Britain.

Mumby, P.J., Green, E.P. Edwards, A.J. and Clark, C.D., 1998a, Coral reef habitat mapping: How much detail can remote sensing provide? *Marine Biology*, **130**, 193–202.

Mumby, P.J., Clark, C.D., Green, E.P. and Edwards, A.J., 1998b, Benefits of water column correction and contextual editing for mapping coral reefs. *International Journal of Remote Sensing*, **19**, 203–210.

Nehls, G. and Thiels, M., 1993, Large-scale distribution patterns of the mussel Mytilus edulis in the Wadden Sea of Schleswig-Holstein: do storms structure the ecosystem? *Netherlands Journal of Sea Research*, **31**, 181–187.

Newbery, D.M., Songwe, N.C. and Chuyong, G.B., 1998, *Phenology and dynamics of an African rainforest at Korup, Cameroon*. In: D.M. Newbery, H.H.T. Prins and N.D. Brown (eds) Dynamics of tropical communities. BES Symp. Vol. 37, 267 – 308. Oxford, Blackwells Scientific.

Norton, T.W. and Williams, J.E., 1992, Habitat modelling and simulation for nature conservation: a need to deal systematically with uncertainty. *Mathematics and Computers in Simulation*, **33**, 379–384.

Norton, T.W. and Possingham, H.P., 1993, *Wildlife modelling for biodiversity conservation. Modelling Change in Environmental Systems*. Edited by Jakeman, A.J., Beck, M.B. and McAleer, M.J.. New York, John Wiley and Sons.

Norton-Griffiths, M., 1978, *Counting Animals*. handbook No.1 in a series of Handbooks on techniques currently used in African wildlife ecology, Grimsdell J.J.R. (ed.), Serengeti Ecological Monitoring programme, African Wildlife Leadership Foundation, Nairobi, Kenya.

Oindo, B., 1998, *Analysis of the transferability of species environment models: A case study of topi in Masai Mara ecosystem, Narok Districk, Kenya*. MSc. Thesis. ITC, Enschede, The Netherlands.

Omullo, L., 1995, *Modelling the distribution of wildlife in relation to environmental gradients. A case study of the Tsavo National Park, Kenya*. MSc-thersis, ITC, Enschede.

Omullo, L., 1996, The utility of biological modelling and GIS in wildlife management: a case study of some mammals in Tsavo National Park. *Proceedings of the conference on the application of remotely sensed data and GIS in environmental and natural resources assessment in Africa*. Harare, Zimbabwe, 15–22 March 1996.

Osborne, P.E. and Tigar, B.J., 1992, Interpreting bird atlas data using logistic models: an example from Lesotho, Southern Africa. *Journal of Applied Ecology*, **29**, 55–62.

Pereira, J.M.C. and Itami, R.M., 1991, GIS-based habitat Modeling Using Logistic Multiple regression: A Study of the Mt. Graham Red Squirrel. *Photogrammetric Engineering and Remote Sensing*, **57**, 1475–1486.

Pomeroy, D. 1993, Centres of high biodiversity in Africa. *Conservation Biology*, **7**, 901–907.

Pope, R.J., Bolog, G.G., Patel, M. and Sokhey, A.S., 1992, Water quality-The back River is back. *Water Environment and technology*, **4**, 32–37.

Prince, S.D., 1991, A model of regional primary production for use with coarse resolution satellite data. *International Journal of Remote Sensing*, **12**, 1313–1330.

Prince, S.D. and Tucker, C.J., 1986, Satellite remote sensing of rangelands in Botswana II. NOAA AVHRR and herbaceous vegetation. *International Journal of Remote Sensin*, **7**, 1555–1570.

Prins, H.H.T., 1987, Nature conservation as an integral part of optimal land use in East Africa: the case of the Masai Ecosystem of northern Tanzania. *Biological Conservation*, **40**, 141-161.

Prins, H.H.T., 1996, *Ecology and Behaviour of the African buffalo: Social inequality and decision-making*. London, Chapman and Hall.

Prins, H.H.T. and Henne, G., 1998, *C.B.D. workshop on the Ecosystem Aproach, Lilongwe, Malawi*. Report to the Permanent Secretary of the Convention on Biological Diversity in Montreal for the Government of The Netherlands and the Government of Malawi. www.biodiv.org\cop5\decisions.

Prins, H.H.T. and Olff, H., 1998, Species richness of African grazer assemblages: towards a functional explanation. In: D.M. Newbery, H.H.T. Prins and N.D. Brown (eds) Dynamics of tropical communities. *BES Symposium*, **37**, 449 – 490. Oxford, Blackwells Scientific.

Prins, H.H.T. and Ydenberg, R.C., 1985, Vegetation growth and a seasonal habitat shift of the barnacle goose (Branta leucopsis). *Oecologia*, **66**, 122-125.

Rappole, J.H., Powell, G.V.N. and Sader, S.A., 1994, *Mapping the Diversity of Nature*. Edited by Ronald I. Miller. London, Chapman and Hall.

Richards, J.A., Woodgate, P.W. and Skidmore, A.K., 1987, An explanation of enhanced radar backscattering from flooded forests. *International Journal of Remote Sensing*, **8**, 1093–1100.

Risser, P.G., Karr, J.R. and Forman, R.T.T., 1984, *Landscape ecology: directions and approaches.* Illinois Natural History Survey Special Publ. Nr. 2. Champaign, Illinois Natural History Survey.

Rodriguez, D.J., 1997, A method to study competition dynamics using de Wit replacement series experiments. *Oikos*, **78**, 411.

Rosenzweig, M.L.,1995, *Species diversity in space and time.* Cambridge, Cambridge University Press.

Said, M.Y., Ottichilo, W.K., Sinange, R.K.and Aligula, H.M., 1997, *Population and distribution trends of wildlife and livestock in the Mara ecosystem and the surrounding areas.* Kenya Wildlife Service (KWS), Nairobi, Kenya.

Skidmore, A.K. and Turner, B.J., 1989, Assessing the Accuracy of resource Inventory maps. In *Proceedings of the "Global natural Resource Monitoring and Assessments: Preparing for the 21th Century*, September 24–30,Venice, Italy, pp. 524–535.

Skidmore, A.K., Gauld, A. and Walker, P., 1996, Classification of kangaroo habitat distribution using three GIS models. *International Journal of Geographical Information Systems*, **10**, 441–454.

Skidmore, A.K., Turner, B.J., Brinkhof, W and Knowles, E., 1997, Performance of a Neural network: Mapping Forest Using GIS and Remotely Sensed Data. *Photogrammetric Engineering and Remote Sensing*, **63**, 501–514.

Skidmore, A.K., 1998, Nonparametric Classifier for GIS data Applied to kangaroo distribution Mapping. *Photogrammetric Engineering and remote sensing*, **64**, 217–226.

Skidmore, A.K. and Kloosterman, H., 1999, *Hyperspectral remote sensing for mapping native vegetation.* Remote sensing nieuwsbrief, **22**, 82.

Skidmore, A.K., 1999, Accuracy assessment of spatial information. In: Stein, A., van der Meer, F. and Gorte, B., 1999. *Spatial Statistics for Remote Sensing.* Kluwer Academic Publishers, Dordrecht, Chapter 12, 197–209.

Stoms, D.M., Davis, F.W. and Cogan, C.B., 1992, Sensitivity of wildlife habitat models to uncertainties in GIS data. *Photogrammetric Engineering and Remote Sensing*, **58**, 843–850.

Soulé, M.E., 1986, *Conservation biology: the science of scarcity and diversity.* Sunderland Massachusetts, Sinauer Assoc.

Thomas, J.A. and Bovee, K.D., 1993, Application and testing of a procedure to evaluate transferability of habitat suitability criteria. *Regulated rivers: Research and management*, **8**, 285–294.

Thouless, C. and Dyer, A., 1992, Radio-tracking of Elephants in Laikipia District, Kenya. *Pachyderm*, **15**, 34–39.

Toxopeus, A.G., 1996, *ISM, an interactive Spatial and temporal Modelling system as a tool in ecosystem management.* PhD-thesis, ITC, Enschede, 250

Toxopeus, A.G., 1998, Spatial and temporal modelling for sustainable management of semi-arid rangelands: the wildlife verrsus livestock issue in the Ambeseli ecosystem, Southern Kenya. *Proceedings of the conference on geo-information for sustainable land management*, ITC, Enschede, The Netherlands.

Toxopeus, A.G. and W. van Wijngaarden, 1995, An interactive spatial modelling (ISM) system for the management of the Cibodas Biosphere Reserve (West Java, Indonesia). *ITC Journal*, **1994–4**, 385–391.

Tucker, C.J., 1979, Red and Infra-red linear combination for monitoring vegetation. *Remote Sensing Environment*, **8**, 127–150.

Tutin, C.E.G. and White, L.J.T., 1998, Primates, phenology and frugivory: present, past, and future patterns in the Lope Reserve, Gabon. In: Newbery, D.M., Prins H.H.T. and Brown, N.D. (ed.) *Dynamics of tropical communities. BES Symposium*, **37**, 309 – 337. Oxford, Blackwells Scientific.

US Fish and Wildlife Service (USFWS), 1981, *Standards for the development of habitat suitability index models.* 103 ESM. U.S. Fish Wildl. Serv., Div. Ecol. Serv., Washington D.C.

Van der Meer, F., 1995, Spectral unmixing of Landsat Thematic Mapper data. *International Journal of remote Sensing*, **16**, 3189–3194.

Van Oene, H., Van Deursen, E.J.M. and Berendse, F., 1999, Plant-herbivore interaction and its consequences for succession in wetland ecosystems: a modelling approach. *Ecosystems*, **2**, 122 – 138.

Van Zyl, J.J., Burnette, C.F. and Farr, T.G., 1991, Inference of surface power spectra from inversion of multifrequency polarimetric radar data. *Geophysical Research Letters*, **18**, 1787–1790.

Varekamp, C., Skidmore, A.K. and Burrough, P.A., 1996, Development of public domain geostatistical software for mapping forest soils. *Photogrammetric Engineering and Remote Sensing*, **62**, 845–854.

Veenendaal, E.M. and Swaine, M.D., 1998, *Limits to tree species distributions in lowland tropical rainforest.* In: Newbery, D.M.. Prins H.H.T and Brown N.D. (ed.) *Dynamics of tropical communities.* BES Symposium 37, 163 – 191. Oxford, Blackwells Scientific.

Verbyla, D.L. and Litvaitis, J.A., 1989, Resampling methods for evaluating classification accuracy of wildlife habitat models. *Environmental Management*, **13**, 783–787.

Walker, P.A., 1990, Modelling Wildlife distributions using a geographic information system: kangaroos in relation to climate. *Journal of Biogeography*, **17**, 279–289.

Walker, P.A. and Moore, D.M., 1988, SIMPLE, an inductive modelling and mapping tool for spatially-oriented data. *International Journal of geographical Information Systems*, **2**, 347–363.

Wamukoya, G.M., Mirangi, J.M., and Ottichilo, W.K., 1995, *Aerial survey of sea turtles and marine mammals.* Technical Report Series No. 1. Kenya Wildlife Service, Nairobi, Kenya.

Waweru, N.M., 1998, *Mapping rangeland quality using remote sensing, environmental data and expert knowledge.* MSc thesis, ITC, Enschede, The Netherlands.

WCMC, 1992, *Global biodiversity: Status of the Earth's living resources.* London, Chapman and Hall.

Weeks, R.I., Smith, M., Kyung Pak, Wen-Hao-Li, Gillespie, A. and Gustafson, B., 1996, Surface roughness, radar backscatter, and visible and near- infrared reflectance in Death Valley, California. *Journal of Geophysical Research*, **101**, 23077–23090.

Western, D., 1989, *Why manage nature?* In Western, D. and Pearl, M.C., (Eds.), Conservation for the twenty-first century. New York, Oxford University Press.

Western, W., and Gichohi, H., 1993, Segregation effects and the impoverishment of savanna parks:the case for ecosystem viability analysis. *African Journal of Ecology*, **31**, 269–281.

Wiersema, G., 1983, Ibex habitat analysis using Landsat imagery. *ITC Journal*, **1983–2**, 139-147.

Williams, G.L., 1988, An assessment of HEP (Habitat Evaluation Procedures) applications to Bureau of Reclamation projects. *Wildlife Society Bulletin*, **16**, 437–447.

Woodworth, B.L., Farm, B.P., Mufungo, C., Borner, M. and Kuwai, J.O., 1997, A photographic census of flamingos in the Rift Valley Lakes of Tanzania. *African Journal of Ecology*, **35**, 326–334.

Wylie, B.K., Harrington, J.A., Prince, S.D., and Denda, I., 1991, Satellite and ground–based pasture production assessment in Niger: 1986–1988. *International Journal of Remote Sensing*, **12**, 1281–1300.

Zagt, R.J. and M.J.A. Werger, 1998, Community structure and the demography of primary species in tropical rainforest. In: Newbery, D.M. Prins, H.H.T. and Brown, N.D. (eds) *Dynamics of tropical communities. BES Symposium*, **37**, 193-219. Oxford, Blackwells Scientific.

Biodiversity mapping and modelling

J.R. Busby

8.1 CONTEXT

The current global decline and loss of biological diversity (biodiversity) is now a major public-policy issue, with a number of international conventions focussing specifically on biodiversity, e.g. *The Convention on Biological Diversity* (CBD), *Convention on International Trade in Endangered Species of Wild Fauna and Flora* (CITES), *Convention on Migratory Species* (Bonn Convention) and *The Ramsar Convention on Wetlands* (see References). Contracting Parties to these conventions, generally nation states, are obligated to report on the status of, and trends in, biodiversity, within their respective jurisdictions.

Increasingly, countries are also recognizing that biodiversity issues are not contained within national borders, so regional initiatives are also under way, such as the *Agreement on the Conservation of Bats in Europe* (EUROBATS 1991). Some 33 international agreements related to the conservation of biological diversity are currently listed in the Environmental Treaties and Resource Indicators (ENTRI) system (CIESIN 1998).

Biodiversity mapping and modelling is becoming increasingly important to the successful implementation of these initiatives, not only to the governments and intergovernmental agencies and programmes directly involved, but also at local levels including individuals and community groups, as well as indigenous peoples.

Agencies and individuals involved in the assessment and management of living resources at all geographic scales need the insights provided by scientific research on the nature and distribution of biodiversity. This chapter focuses on the mobilization of spatial data, as well as the use of tools and techniques to generate information from these spatial data, that informs various stakeholder groups on the many dimensions of biodiversity.

8.2 DEFINITIONS

Article 2 of the *Convention on Biological Diversity* specifies the following 'Use of Terms' for the purposes of the Convention:

- **'Biological diversity'** means the variability among living organisms from all sources including, *inter alia*, terrestrial, marine and other aquatic ecosystems and the ecological complexes of which they are part; this includes diversity within species, between species and of ecosystems.
- **'Ecosystem'** means a dynamic complex of plant, animal and micro-organism communities and their non-living environment interacting as a functional unit.

8.3   KEY ISSUES

There are fundamental challenges in translating the above definitions into operational programs. These centre around the question 'What is a high biodiversity value?'. For example, a high biodiversity value may be attached to a group (e.g. family) in which a large number of species have been described, yet a single species in an isolated region may possess more genetic diversity than these combined species. Similarly species-rich ecosystems, such as tropical rainforests, tend to attract more attention than species-poor ecosystems occurring in harsher environments, such as arid and semi-arid regions. However the latter species may be of very high evolutionary, ecological, social or economic value. All these issues need to be balanced when scarce resources are being allocated. The key point here is that the value of the answer depends very much on the question being asked. The challenge is to ask the right question, such as 'what is a high biological diversity value?'.

Any framework for biodiversity mapping and modelling must, *inter alia,* bridge the gap between science-based data on the nature and distribution of biological diversity and knowledge that is relevant to people, in particular policy makers and decision takers. The following questions exemplify this transition, going from reasonably straightforward baseline inventory to complex scenario building, whereby a resource manager can not only explore various options for action in the field but also provide policy-relevant advice to executive management:

- what are the various elements of biodiversity?
- where does something (species, ecosystem) occur?
- what (species of interest, some environmental resource) is found in a particular place (protected area, administrative zone)?
- what (environments, ecosystems, some environmental resource) exist and where are they found?
- how are environments being managed?
- is something (species, ecosystem, some environmental resource) changing, by how much, is it important, what can be done about it?
- what will happen if a perturbation (fire, global warming, agricultural activity etc.) is made to the ecosystem?

Note that the scientific issues that dominate the earlier questions are progressively counterbalanced by social and economic issues in later ones. Many tools have been developed and much data have been collected in attempts to answer these questions. Regrettably, invalid assumptions, inadequate or inaccurate data and flawed procedures have rendered many of these efforts little more than 'computer games', of very little practical benefit to the assessment and management of living resources.

8.4 MOBILIZING THE DATA

Contrary to popular belief, there are masses of data 'out there'. They are scattered, incomplete, of variable quality and poorly documented. In too many cases, it has proven cost-effective to collect new data rather than to attempt to collate and upgrade existing data (See also Chapter 3 and 4).

8.4.1 Attribute selection

The first decision in mobilizing the data is about what data to collect. The attribute selection prior to data collection will dominate future options for modelling and interpretation. Attributes can be 'factual' or 'primary' or, alternatively, 'derived' or 'classified'. Examples of primary attributes include latitude and longitude of the place where an observation was made, date of that observation, height of a tree, or mean annual temperature of a site. These data can all be measured or otherwise described against a stable, objective or widely accepted standard.

People are, however, uncomfortable with working with raw, unprocessed data. They instinctively like to classify data into categories that have greater meaning to them. Such derived attributes are those developed from primary attributes through a process of interpretation or classification applied at the time, or subsequently, according to some paradigm. These include: species name, soil type, vegetation class, climate zone.

The difficulty is that these classified categories are erected for a specific purpose and are not necessarily widely shared or understood. Thus other people may have great difficulty in understanding what these categories mean. In general, therefore, derived attributes should not be recorded in a database unless the primary attributes from which they were derived are also available. This is because, as concepts and paradigms change, derived data degrade in value and may even become useless. For example, if the only representation of a species distribution is a map stored as a polygon coverage in a geographic information system (GIS), this distribution is valueless if that species is split following a taxonomic revision. Examples of primary and derived attributes are shown in Table 8.1.

Table 8.1: Examples of primary vs. derived attributes.

Primary	Derived
Geocode (i.e. a point reference such as latitude and longitude)	Grid cells (e.g. 1 km grids), administrative zones (e.g. counties)
Plant height in absolute units, e.g. metres	Plant height as 'tall', 'medium' or 'low'
Actual time of observation	'Early morning', 'dusk'
Actual date	'Summer'
Start and end dates of observation	
Mean annual temperature	'Hot'

A particularly difficult example is species name, a derived attribute for which there is no feasible alternative. Most researchers have no option but to use species names (where there is one), even though these change from time to time. It is simply not practicable to store all the myriad primary attributes of an organism so that it can be unambiguously allocated to a taxon whenever its group is revised.

The principle is clear: we need to minimise the risk of datasets becoming obsolete as a result of changing concepts and classifications.

8.4.2 Sampling design

The way in which records are made in the field fundamentally determines the potential use for the data. Decisions made about sampling strategies will influence options of which analytical tools can be used.

The selection of a study area is the first important decision. In some cases the problem determines the area, e.g. for a project investigating the distribution of lizard species on a (small) island, the study area is self-evident. However, more commonly, the study area is selected from some larger region. The selection may be made according to some probability sampling scheme, or it may simply reflect the recorder's view that the study area is in some sense representative of the larger region. However it is done, data collection needs to be 'representative' of the environmental feature(s) under investigation in order for the analysis to lead to reliable and useful conclusions. If the data are to be quantitatively analyzed, the statistics may be based on certain assumptions about the sampling scheme, such as sample selection is random and independent. To ensure representativeness, sampling in the field may be stratified (Cochran 1977). Various sampling schemes (e.g., random, stratified random, regular) may be used depending on the issue under consideration, the nature of the environment, logistics, and the proposed analysis tools, and are reviewed in Cochran (1977).

One of the objectives of field data collection is to distinguish patterns in the distribution of biodiversity, identify their possible causes, and predict future behaviour. Thus, biologists are not so much interested in showing that an observed pattern departs significantly from 'complete spatial randomness', as in interpreting and understanding the pattern (Diggle 1983).

8.4.3 Data capture

Ideally, the individual attributes should be recorded in compliance with some standard but, even more importantly, in a consistent way. Where there are competing standards, or no standard that is either available or followed, it is vital to be consistent and to thoroughly document the actions taken in the capture process. It is generally far easier to convert a dataset that is recorded consistently, even though not in compliance with a standard(s), than to convert a dataset that purports to meet a standard, but is inconsistent. A framework for standards applicable to specimens and observations of species collected in the field is shown in Table 8.2. Note that missing (e.g. impossible to measure or absent) attribute values should be indicated as such, rather than the fields just left blank.

8.4.4 Standards and quality assurance

Data collection is expensive, thus it is important to maximize the use of data. The collection and management of data with multiple uses in mind will bring the highest return on investment. This is aided significantly by the adoption and implementation of standards for recording and managing biodiversity data and by quality-assurance procedures that ensure data meets the required standard(s). An example of standards is given in Table 8.2.

Table 8.2: Indicative attributes and standards for species occurrence records.

Attribute	Standard	Notes
Record class	Type of record	Specimen – should include information on the collection (e.g. museum name or identifier) and collection identifier
		Observation – should include name of observer
		Literature – should include (link to) full bibliographic reference
Taxon name	Taxon authority list	National checklist (e.g. Australia : Census of Australian Vertebrate Species (CAVS) Version 8.1) international list (e.g. Species 2000). May need to include supra- and infra-specific names to provide further context or additional detail, or even broader categories where the taxon cannot be identified with precision
Georeference (geocode)	Latitude and longitude	Universally applicable but needs to be recorded consistently (e.g. degrees, minutes, seconds)
	Map grid reference	Depends on the mapping system and can be very difficult to convert to other systems, such as latitude-longitude. The full reference needs to be recorded, rather than abbreviated ones and full details of the map coordinate system, including origin references, should be recorded
	Customized grid	These are frequently developed to meet particular project objectives (e.g. publication of distribution maps at particular scales), but can be extremely difficult to convert. Often the grids are so large that valuable details of locality are lost
Locality	Named place from gazetteer, often qualified by distance and direction from named place	Name may need to be further qualified if this is not unique, e.g. there are many different 'Sandy Creek's, in the Australian national gazetteer
Date and time	Year/month/day hour:minutes	May need to accommodate ranges for extended observation periods

[Note that other secondary supporting attributes will also be required, e.g. collector(s) name, identifier(s) name, date of identification, altitude, depth, previous names applied to this record, geographic region (e.g. catchment), administrative region (e.g. county – remember that these boundaries can change over time), qualifiers on any of the above, e.g. geocode precision].

The purpose of standards is to *minimise the transaction costs of using data.* They are the means to expedite information communication amongst people, from different disciplines, who examine an environmental issue from different perspectives. Standards cover:
- the selection of attributes representing the environmental feature(s) under investigation
- data collection methods and survey protocols
- the meaning of those attributes (allowable numeric ranges, values, etc.)
- methods of documenting (metadata) and assuring the quality of those attributes: their representation (spatial, tabular, text, etc.), management, security, etc.
- how those attributes are communicated to others via various media.

Standards for generation, management and quality assurance of biodiversity data are very difficult to establish because of the wide variation in species attributes and, consequently, of work practices of institutions that handle them (see Box 8.1). Some examples of standards are listed in Box 8.2.

Box 8.1: Museum collections databases

Extract from *Statement of the Problem* from the Oz Project at the University of Kansas

Unfortunately, there are no standards for computerised collection management. Most institutions have developed unique, in-house solutions to handle their computational needs. Even within the same museum, different collections often use different database designs and database management software. This makes it difficult, if not impossible, to access information simultaneously about the holdings in multiple collections or institutions. This practice is also costly and difficult for an institution to maintain because expertise in many database designs and RDBMS software is needed.

There are many reasons for this situation. The most important is that different disciplines maintain different types of data. For example, the information stored about specimen measurements for a mammal may be completely different than for a fish. Habitat and preparation data are other important examples. In addition, certain types of data may be more important for some collections than it is for others. Entomologists, for instance, are much more concerned with easily changing taxonomic identifications of specimens than ornithologists or mammalogists. This means even when different collections maintain the same kind of data, the systems they use to access the data may differ functionally.

Another cause of divergent functionality is that different collections often use dissimilar management practices. The process of collecting, accessioning, preparing, identifying and cataloguing a specimen in a herpetology collection may be completely different than that for a specimen in an entomology collection, and the software developed for the respective collections often reflects this. Also, collection management systems have mainly been developed to support specimen-based collections. However, there are non-specimen based collection items which need record keeping.

Box 8.2: Biodiversity standards and protocols (US)

National Survey of Land Cover Mapping Protocols Used in the Gap Analysis Program
<http://www.calmit.unl.edu/gapmap/report.htmi>

Methods for Assessing Accuracy of Animal Distribution Maps, Blair Csuti and Patrick Crist
<http://www.gap.uidaho.edu/handbook/VertebrateDistributionAssessment/default.htm>

Methods for Developing Terrestrial Vertebrate Distribution Maps for GAP Analysis, Blair Csuti and Patrick Crist
<http://www.gap.uidaho.edu/handbook/vertebrateDistributionModeling/default.htm>

Vegetation Classification Standard, Vegetation Subcommittee, Federal Geographic Data Committee, June 1997 <http:://www.fgdc.gov/standards/documents/standards/vegetation>

8.4.5 Data custodianship and access

Field collection of data is desirable because it is both current and the attributes recorded can be tailored to the purpose of the study. However, for cost and other reasons, this may not always be possible and use must be made of existing data, at least in part. Existing data can be extremely useful in extending or interpolating field observations and, of course, there is little alternative for studies of changes over long periods of time.

The process of finding other data is aided considerably by the progressive development of clearing houses and metadatabases (see examples in References and Chapter 4). Such catalogues can be very useful in locating potentially-useful datasets.

Existing data are, of course, generally managed by others. These 'data custodians' have very varied approaches to data management and equally varied policies and procedures relating to data access. Custodians have a number of specific concerns about releasing data, including:

- will the dataset be used 'correctly'?
- is the exchange consistent with corporate policy?
- will use of the dataset be fully acknowledged?
- could credibility suffer (e.g. where data are found to be of poor quality)?
- will costs be covered?
- is there any vulnerability to legal liability in the event that the data are shown to be incorrect and some harm has resulted?

These concerns are most effectively addressed in a data access agreement between the data custodian and the potential user. Data access agreements can be quite varied, but usually include the following elements:

- permitted/excluded uses
- how/whether to distribute to third parties (normally not permitted)
- how to acknowledge
- details of any transaction costs
- a disclaimer (to protect the custodian from legal liability).

Normally different provisions (especially transaction costs) vary with the class of user. Scientific and public education uses are generally less restricted than commercial uses, for example. Normally it is strongly advised to fulsomely acknowledge the sources of data used, regardless whether this is a provision of the data access agreement. This builds a level of trust between custodians and users that is essential to the unrestricted flow and multiple use of data.

8.4.6 Data mining and harmonization

Although considerable volumes of data may exist, they are all too often scattered, incomplete and of undocumented quality. Nevertheless, scarce resources can often be cost-effectively employed in extracting data from one or more existing sources ('data mining') and, as necessary, merging them ('data harmonization').

Once a potential dataset has been located, further investigation will determine whether any potentially useful records are present. Once these have been extracted, they will generally need to be 'harmonized' with other records. This process is considerably aided if all the records meet the same or closely related standards or, failing that, if they have been recorded consistently within their home dataset. In the latter case, it is often possible to convert the data to a standard form by automated procedures. Otherwise it can be a very labour-intensive process to convert large number of records on a case-by-case basis.

Of course, once a composite dataset has been laboriously constructed, it is important to document the dataset and the process by which it was built, in order to ensure that the dataset becomes of maximum value to other projects.

8.5   TOOLS AND TECHNIQUES

While collection of large volumes of data is relatively straightforward, these raw data are generally not particularly useful without further analysis and interpretation. Understanding is enhanced when data are converted into information through the application of appropriate tools and techniques. Different kinds of tools are needed for different stages in the 'information manufacturing' process as raw data is progressively converted to useful information products up the 'information pyramid' (Figure 8.1).

There are many tools currently available and more are being developed all the time. These tools vary considerably, not only in the tasks they perform but also in reliability, documentation and support. Tools range from commercial products from major software suppliers, through packages that have been developed and are supported to varying extents by government agencies, scientific institutions and universities, through to those developed by individuals for their own purposes. The last of these vary considerably in quality. They may be very difficult to load and operate, they may be unreliable and they are often poorly documented and unsupported.

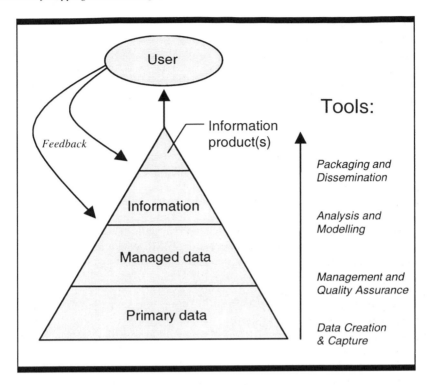

Figure 8.1: Tools to support data flow up the 'information pyramid'
(modified from Figure 2 in World Conservation Monitoring Centre (1998a)).

Detailed evaluation of data generation, management and quality assurance tools is beyond the scope of this chapter. However it is important to ensure that correct choices are made among those available in order to expedite analysis and modelling. As indicated above, choices made with data collection and selection will inevitably influence options for analysis and interpretation. All too often, major deficiencies in the data are obscured by subsequent analyses, resulting in maps and other products that, although visually impressive, are very misleading. Worse still, they could even result in the misallocation of scarce resources or unintended environmental outcomes.

It is also vital to understand the audience or intended target for the results. An impressive, but highly complex model will baffle most senior managers and other non-specialists. The data may be impeccable and the science world class, but managers will be very reluctant to accept the results and will seize on almost any excuse to avoid doing so. On the other hand, a very simple analysis may be summarily dismissed as not dealing adequately with the complexity of the 'real world'.

8.5.1 Database management

Most species occurrence data is managed as sets of 'tables' (files) by 'relational database management systems' (RDBMS). Each table is an 'entity' (such as a species) which possesses a number of 'attributes', such as name, life form, conservation status, etc. These entities are linked by a set of 'relationships', for example a 'species' entity is linked to a 'site' entity by a relationship that a species has been recorded at a site. Entity-Relationship (E-R) diagrams are frequently used to document these. RDBMS software manages the entities and relationships, allowing data entry, updating, searching, sorting and reporting.

Associated with RDBMS has been the progressive development of 'Structured Query Language' (SQL). This is an English-like retrieval language for querying relational databases. SQL is database independent, which frees organizations from being committed to any particular brand of RDBMS. SQL has also benefited from the development of client-server database architectures. Client-server is currently built into most commercial RDBMS packages, which facilitates the development of datasets that can be accessed by a wide range of users in remote locations (Olivieri *et al.* 1995*)*.

8.5.2 Geographic information systems

GIS, and its role in environmental modelling, is covered throughout this book. In the following section, some of the aspects of raster and vector data as they apply to biodiversity data will be outlined.

8.5.2.1   *Raster data*

Raster data structure is an abstraction of the real world where spatial data are expressed as a matrix of cells or pixels, with spatial position implicit in the ordering of the pixels. With the raster data model, spatial data are not continuous but divided into discrete units. This makes raster data particularly suitable for certain types of spatial operation, for example overlays, area calculations, or simulation modelling, where the various attributes for each pixel can be readily manipulated because they are referenced to a common geographic base. Unlike vector data however, there are no implicit topological relationships. Remote sensing data are largely stored and manipulated in raster form.

With biodiversity data, a raster format can be particularly effective for displaying results, and especially for biodiversity units that cover substantial areas, such as large vegetation units, or modelling predicted distributions of species. A raster format is not recommended for storing input data, unless the pixels are extremely small, i.e. equivalent to points, or otherwise very much smaller than the scale at which analyses will be required. This is because of difficulties in disaggregating data for areas that are subsets of individual pixels or that cross pixel boundaries. In such cases it is uncertain whether or not the pixel attributes apply to the areas under investigation, and the data may therefore be unusable.

8.5.2.2 *Vector data*

The vector data structure is an abstraction of the real world where positional data is represented in the form of coordinates. In vector data, the basic units of spatial information are points, lines and polygons. Each of these units is composed simply as a series of one or more coordinate points, for example, a line is a collection of related points, and a polygon is a collection of related lines. Typical vector data include administrative boundaries (polygons), road networks (lines) and sites where rare species have been recorded (points).

Points are commonly used to represent individual records of species, although polygons are also used to represent species distributions and vegetation and environmental units. Comparatively few biodiversity attributes are linear, although some vegetation types, such as mangroves or riparian vegetation, may be quasi-linear, or be represented that way at coarse scales.

As indicated above under *Attribute selection,* polygons should not be used for storing raw data on species distributions because of the difficulties of disaggregating them if identifications change. However they are very effective for display purposes.

8.5.3 Distribution mapping tools

8.5.3.1 *Where does a species occur?*
Hand-drawn maps or simple GIS are all that is required to answer this question, given appropriate raw data. The same tools can be used to plot the locations of ecosystems. More sophisticated tools can deliver dynamic maps over the Internet. For example the service once provided by Environment Australia prompted for a species name then, once the user had made a selection, initiated a SQL query on a relational database, put the retrieved data through a mapping tool and returned a map with associated metadata (Figure 8.2).

BIOCLIM is one example of a tool that uses environmental parameters, in this case climate, to estimate species distributions (Box 8.3). The above species mapping service provided BIOCLIM predictions on-line.

Box 8.3: BIOCLIM

BIOCLIM can be used to analyse and estimate the distribution of any entity – animal or plant species or vegetation type – that is influenced by climate. BIOCLIM requires climate surfaces that are used to produce site-specific estimates of monthly temperature and precipitation for places where the entity has been recorded. The climate estimates are then aggregated into a 'climate profile' (see Chapter 2), using parameters that are indicative of annual mean, seasonal and monthly extreme values.

Predicted distributions are based on the similarity of climates at points on some geographic grid to the climate profile.

BIOCLIM has been used to model the distributions of a wide variety of organisms, including temperate rainforest trees (Busby 1986), snakes (Longmore 1986), bats (e.g. Walton *et al.* 1992) and brine shrimps (Williams and Busby 1991). It can also reconstruct palaeohistoric distributions (e.g. McKenzie and Busby 1992, Kershaw *et al.* 1994) and predict the potential impacts of climate change (e.g. Busby 1988; Dexter *et al.* 1995).
[Source: Busby (1991)]

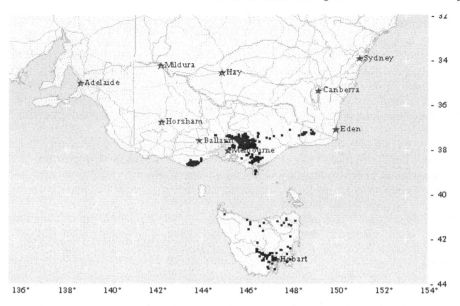

Data Source/ Collection Information for *Eucalyptus regnans*		
Institution	Collection Dates	Number of Records
Australian National Botanic Gardens Herbarium, Canberra	1969–1983	10
Australian National Herbarium, CSIRO, Canberra	1920–1986	50
CSIRO Tree Seed Centre, Canberra	1962–1989	77
Department of Conservation and Natural Resources, Victoria	1900–1990	1,228
NSW Herbarium, Sydney	1899–1986	40
National Herbarium of Victoria, Melbourne	1867–1991	55
Queensland Herbarium, Brisbane	1955–1976	5
Tasmanian Herbarium, Hobart	1897–1980	51
Total: 1,516		

Figure 8.2: Distribution of *Eucalyptus regnans*, the world's tallest hardwood tree species
Source: **The data, from the custodians listed above, as in the ERIN database at 25 March 1999.
The base map is based on spatial data available from the Australian Surveying and Land
Information Group (AUSLIG).**

8.5.3.2 What is found in a particular place?

The answer to this question is somewhat more complex, depending on how that 'place' is defined relative to the way the data are stored. If the data are stored as layers in a GIS and the place is a defined polygon (or its raster equivalent, if appropriate), then the required layers can be overlaid and the polygon of interest clipped out, a process known as 'cookie cutting'. Similarly, if the place of interest is a defined attribute of data stored in a RDBMS, then a simple retrieval of all records possessing that attribute will achieve the desired result.

On the other hand, if the place is a polygon in a GIS and the data are stored as tables in a relational database, or the data are GIS layers but the place has no defined spatial attributes, then the task can be very difficult.

8.5.4 Environmental domain analysis

8.5.4.1 What environments exist and where are they found?

Classifying a large area into a number of regions that are relatively homogeneous for some set of environmental attributes provides a useful framework for focussing attention, summarizing patterns, aggregating information, and allocating resources and priorities in nature conservation. These regions can be interpreted as vegetation types, ecosystems, landscapes, eco-regions, biomes or environmental domains, depending on the attributes chosen, the map scale, and the objectives of the analysis. Environmental attributes can include climate, lithology/geology, landform, soils, vegetation, flora and fauna, and land use (Box 8.4).

Box 8.4: Interim Biogeographic Regionalization of Australia

The Interim Biogeographic Regionalization of Australia (IBRA) is an integrated classification of both biotic and abiotic variation. IBRA regions represent a landscape-based approach to classifying the land surface, including attributes of climate, geomorphology, landform, lithology, and characteristic flora and fauna. The developers acknowledge that new information will modify our understanding of the regions, hence the term interim (Thackway and Cresswell 1995).

The National Reserves System Cooperative Program (NRSCP) needed a classification of ecosystems agreed by all Australian nature conservation agencies. The IBRA was developed to provide an ecological framework within which to identify gaps in the national reserves system and to set priorities for gap filling.

By itself the IBRA is of little value in assisting decision makers determine gaps and set priorities. The value of IBRA for this purpose lies in the development of conservation planning attributes for each IBRA region. Conservation planning requires details of the following attributes for each region and for sub-regions therein:
1.   key conservation values
2.   reservation status
3.   deficiencies within the existing system of protected areas
4.   types of threats; and
5.   alternative conservation management measures.
Source: <http://www.ea.gov.au/parks/nrs/ibraimcr.index.html>

The underlying premise for such an approach is that physical environmental processes drive ecological processes, which in turn are responsible for the observed patterns of biological productivity and associated patterns of biodiversity. Specialist ecological knowledge is then combined with appropriate biophysical data sets to describe these patterns. The resulting environmental units can be used
for a variety of assessment and planning purposes (Box 8.5).

Box 8.5: Gap Analysis Program, USA

The mission of the Gap Analysis Program (GAP) is to provide regional assessments of the conservation status of native vertebrate species and natural land cover types and to facilitate the application of this information to land management activities. This is accomplished through the following five objectives:

- map the land cover of the US
- map predicted distributions of vertebrate species for the US
- document the representation of vertebrate species and land cover types in areas managed for the long-term maintenance of biodiversity
- provide this information to the public and those entities charged with land use research, policy, planning and management
- build institutional cooperation in the application of this information to state and regional management activities.

It is a cooperative effort among regional, state and federal agencies, and private groups. The purpose of the GAP is to provide broad geographic information on the status of ordinary species (those not threatened with extinction or naturally rare) and their habitats in order to provide land managers, planners, scientists and policy makers with the information they need to make better-informed decisions (Scott and Jennings 1997).

Maps and datasets can be downloaded from the various custodians accessible through the main web site at <http://www.gap.uidaho.edu/default.htm>

8.5.5 Environmental assessment and decision support

The answers to the following questions, i.e.:

- how are environments being managed?
- is something [species, ecosystem, some environmental resource] changing, by how much, is it important, what can be done about it?
- what will happen if …?

are much more complex. Various attempts have been made to develop ecosystem simulation models, expert systems and decision support systems to address these and related issues. Some of these have been outlined in Chapter 2. It is perhaps fair to say that the complexity both of biodiversity itself and in its environmental interrelations has confounded attempts to come up with widely applicable systems. Some very successful systems have been developed for certain areas or ecosystems, but these have proved to be of limited applicability outside the domains in which they were developed.

8.6 DISPLAY AND COMMUNICATION

Many decision-makers, such as civil servants, company directors, local government officials and individual resource users, are too busy or lack the technical background to process large amounts of data or apply themselves to difficult interpretation tasks. They need brief summaries of complex issues, presented in such a way that they can be absorbed quickly without the need for special tools or expertise. **Timeliness** is also a critical factor in determining whether information will be effective at supporting decisions. The most salient aspects of a decision may not be taken into account if key information is not available at the right time (World Conservation Monitoring Centre 1998a).

By emphasizing presentation issues such as clarity, timing and method of delivery, information can be made **useful and usable** by its intended audience. The aim is to take account of the constraints under which decision-makers work, and tailor the information accordingly. The results are often referred to as **information products** rather than sources, reinforcing the idea that they are produced with a specific purpose and user in mind (products which are delivered on a regular basis, perhaps via established procedures and mechanisms, are known as **information services**).

Issues in communicating and disseminating results include:

- who is the audience (policy makers, resource managers, other scientists, civil society)?
- what are their interests and capacities to absorb information, i.e. what will they best respond to?
- what is the most appropriate scale and resolution, both to accurately represent the underlying data and to have maximum relevance to the issue under consideration?
- what is the most appropriate display format(s) – reports, charts, maps, on-line scenario analysis?
- what are the most appropriate dissemination technologies – paper, CD-ROM, Internet?

In their original form, scientific research results are notoriously inaccessible to many, due to their level of complexity, sheer volume and focus on scientific rather than policy issues. This is understandable when the results may not have been intended for use in policy-making, for resource management or by the general public. Nevertheless, scientific information could be of much greater value if it is presented in more appropriate ways.

Techniques for translating scientific understanding into 'policy-relevant' information for decision-makers are currently very poorly developed. This is partly due to the traditional 'stand off' between the scientific and policy-making communities, fuelled by mutual suspicion of each other's goals and methods, which leads to sentiments such as 'the government never use my data' (scientist) or 'the information was too complicated' (government).

8.7   FUTURE DEVELOPMENTS

Technology is increasingly fading as a constraint on developing our understanding of biodiversity issues(see also Chapter 12). The technologies, both information technology and the analysis, modelling and dissemination tools already available are well beyond the grasp of many scientists, let alone senior decision-makers and the wider public. There are loads of data and many knowledgeable people, both scientists and others. The challenges are in mobilizing what we already know and making it more widely accessible and in identifying and filling gaps in our knowledge. This requires a focus on more interdisciplinary work, the building of collaborative data and information exchange networks, and on improved communications.

The major current and future challenges are largely therefore organizational and people, followed by data documentation, data comprehensiveness and quality.

8.8   REFERENCES AND INFORMATION RESOURCES

Agreement on the Conservation of Bats in Europe, EUROBATS, 1991, (http://www.eurobats.org).
Busby, J.R., 1986, A biogeoclimatic analysis of Nothofagus cunninghamii (Hook.) Oerst. in Southern Australia. *Australian Journal of Ecology*, **11**, 1–7
Busby, J.R., 1988, Potential impacts of climate change on Australia's flora and fauna. 387–398 in G.I. Pearman (ed.) *Greenhouse: Planning for Climate Change*. Melbourne, Australia, CSIRO.
Busby, J.R. 1991, BIOCLIM – A Bioclimatic Analysis and Prediction System. 64 – 68 in Margules, C.R. and Austin, M.P. (eds.), *Nature Conservation: Cost Effective Biological Surveys and Data Analysis*. Melbourne, Australia, CSIRO.
Census of Australian Vertebrate Species (CAVS), Version 8.1, <http://www.ea.gov.au/biodiversity/abrs/abif/fauna.html>.
Center for International Earth Science Information Network (CIESIN), 1998, Environmental Treaties and Resource Indicators (ENTRI) [online]. Palisades, New York, CIESIN <http://sedac.ciesen.org/entri>.
Chapman, A.D. and Busby, J.R., 1994, Linking plant species information to continental biodiversity inventory, climate modeling and environmental monitoring, In: Miller, R.I. (ed.), *Mapping the Diversity of Nature*. London, Chapman and Hall.
Cochran, W.G., 1977, *Sampling Techniques*, New York, Wiley.
Convention on Biological Diversity. <http://www.biodiv.org>.
Convention on Biological Diversity, Clearing-House Mechanism, <http://www.biodiv.org/chm>.
Convention on International Trade in Endangered Species of Wild Fauna and Flora <http://www.cites.org>

Convention on Migratory Species <http://www.wcmc.org.uk/cms>

Dexter, E.M., Chapman, A.D. and Busby, J.R., 1995, The Impact of Global Warning on the Distribution of Threatened Vertebrates, ANZECC 1991, Environment Australia Report (unpublished). Canberra, Australia, Environment Australia.

Diggle, P.J., 1983, Statistical Analysis of Spatial Point Patterns, London, Academic Press.

European Topic Centre on Catalogue of Data Sources, <http://www.mu.nieder-sacsen.de/system/cds>.

Gap Analysis Program, USA, <http://www.gap.uidaho.edu/default.htm/> .

Geospatial Data Clearinghouse, US, <http://clearinghouse2.fgdc.gov>.

Joint Website of the Biodiversity-Related Conventions <http://www.biodiv.org/convention/partners-websites.asp>.

Jones, P.G., Beebe, S.E., Tohme, J. and Galwey, N.W., 1997, The use of geographical information systems in biodiversity exploration and conservation. *Biodiversity and Conservation*, **6**, 947–958.

Kershaw, A.P., Bulman, D. and Busby, J.R., 1994, An examination of modern and pre-European settlement pollen samples from southeastern Australia – assessment of their application to quantitative reconstruction of past vegetation and climate, *Rev. Palaeobotany and Palynology*, **82**, 83–96.

Longmore, R. (ed.), 1986, *Atlas of Elapid Snakes of Australia. Australian Flora and Fauna Series Number 7*. Canberra: Aust. Govt. Publ. Service.

McKenzie, G.M. and Busby, J.R., 1992, A quantitative estimate of Holocene climate using a bioclimate profile of *Nothafagus cunninghamii* (Hook.) *Oerst. J. Biogeogr.*, **19**, 531–540.

National Biological Information Infrastructure (NBII) Clearninghouse <http://www.nbii.gov/datainfo/metadata/clearinghouse>.

Olivieri, S.T., Harrison, J. and Busby, J.R., 1995, Data and Information Management and Communication, 607–670 in Heywood, V.H. (ed.), *Global Biodiversity Assessment*. Cambridge, Cambridge University Press.

Scott, J.M. and Jennings, M.D., 1997, A Description of the National Gap Analysis Program, <http://www.uidaho.edu/about/overview/GapDescription/default.htm>.

Socioeconomic Data and Applications Center, Data and Information Catalog Services <http://www.sedac.ciesen.org>.

Species 2000, <http://www.sp2000.org>.

Thackway, R. and Cresswell, I.D., 1995, An Interim Biogeographic Regionalisation for Australia: a framework for establishing the national system of reserves, Version 4.0, Australian Nature Conservation Agency, Canberra, Australia. <http://www.ea.gov.au/parks/nrs/ibraimer/index.html>.

The Ramsar Convention on Wetlands, <http://www.ramsar.org>.

Walton, D.W., Busby, J.R. and Woodside, D.P., 1992, Recorded and predicted distribution of the Golden-tipped Bat *Phoniscus papuensis* (Dobson, 1878) in Australia. *Australian Zoologist*, **28**, 52–54.

Williams, W.D. and Busby, J.R., 1991, The geographical distribution of Triops australiensis (Crustacea: Notostraca) in Australia: A biogeoclimatic analysis. *Hydrobiologia*, **212**, 235–240.

World Conservation Monitoring Centre, 1998a, WCMC Handbooks on Biodiversity Information Management. *Volume 3: Information Product Design.* In: Reynolds, J.H. (Series Editor). London, Commonwealth Secretariat.

World Conservation Monitoring Centre, 1998b, WCMC Handbooks on Biodiversity Information Management. *Volume 7: Data Management Fundamentals.* In: Reynolds, J.H. (Series Editor). London, Commonwealth Secretariat.

8.9 TOOLS AND TECHNOLOGIES

The following list is indicative only. No independent assessment has been made nor is any endorsement expressed. Persons considering using any of these tools should make their own inquiries and assessments.

Name	Type	WWW URL
	Ecosystem process modelling	http://www.nmw.ac.uk/ite/edin/ecos.html
	A one-dimensional upper ocean ecosystem model developed for the central and eastern Pacific Ocean	HTTP://ATHENA.UMEOCE.MAINE.EDU/1DECO-NEW/1DECO.HTM
Alice	Data management	http://dspace.dial.pipex.com/alice/
ANUCLIM	Model climate variables, bioclimatic parameters, and indices relating to crop growth	http://cres.anu.edu.au/software/anuclimtxt.html
BCD	Biological and Conservation Data System	http://www.consci.tnc.org/src/bcdover.html
BG-BASE	Collections management	http://www.rbge.org.uk/BG-BASE/welcome.htm
BG-RECORDER	Plant records management for botanic gardens	http://www.rbgkew.org.uk/BGCI/database.htm
BIOCLIM	Model bioclimatic distributions	SEE ANUCLIM EXAMPLE: http://www.environment.gov.au/search/mapper.html
BioLink	Data and collections management and analysis	http://www.ento.csiro.au/biolink/biolink.html
Biota	Data and collections management	http://viceroy.eeb.uconn.edu/biota
BioTrack	Data management (uses *Biota*)	http://www.bio.mq.edu.au/kcbb/biotrack/biotrack/default.html
BRAHMS	Botanical Research And Herbarium Management System	http://www.camel.co.uk/brahms/

Name	Type	WWW URL
Carto Fauna-Flora	Mapping software to represent animals and/or plant distributions	http://panoramix.umh.ac.be/zoologie/cff/cff_en.html
CASSIA	Collections and Specimen System for Information and Analysis	http://www.nybg.org/bsci/cass/spec.html
CENTURY	A general model of plant-soil nutrient cycling	http://nrel.colostate.edu/PROGRAMS/MODELING/CENTURY/C_main.htm
CLIMEX	Predicting the potential distribution and relative abundance of species in relation to climate	http://www.ento.csiro.au/research/pestmgmt/climex/climex.htm
Condor	Planning tool that integrates biodiversity, social, and economic variables	http://www.conservation.org/SCIENCE/CPTC/INFOTOOL/Condor1.htm
DELTA	DEscription Language for Taxonomy	http://www.keil.ukans.edu/delta/
DYMEX	Population modelling	http://www.ento.csiro.au/research/pestmgmt/dymex/dymexfr.htm
FEDMOD	Forest Ecosystem Dynamics Modeling Environment	http://www.ncgia.ucsb.edu/conf/SANTA_FE_CD-ROM/sf_papers/knox_robert/paper.html
FORMIX3	Rain Forest Simulation Model	HTTP://WWW.USF.UNI-KASSEL.DE/MODELLE/FORMIX.HTM
GARP	Designed for predicting the potential distribution of biological entities from raster based environmental and biological data	http://biodi.sdsc.edu/Doc/GARP/Manual/manual.html
Habitat Suitability Index	Uses mathematical models to compute an HSI value for selected species from field measurements of habitat variables	http://www.mesc.nbs.gov/hsi/hsi.html
Linnaeus II	Biodiversity documentation and species identification	http://www.eti.uva.nl/Products/intro_linn.html
MEKA	Identification of biological specimens	http://www.mip.berkeley.edu/meka/meka.html

Name	Type	WWW URL
Orde	Management of data on the ecology and distribution of insects	http://www.cs.uu.nl/people/jeroen/orde/hl.html
PANDORA	Biodiversity research projects	http://www.rbge.org.uk/research/pandora.home
Platypus	Management of taxonomic, geographic, ecological and bibliographic information	http://www.ento.csiro.au/platypus/platypus.html
PRISMA	Publish databases, generate reports and training materials, and provide information about ecosystems	http://www.conservation.org/SCIENCE/CPTC/INFOTOOL/Prisma1.htm
SimCoast	A fuzzy logic rule-based expert system for analysis of information collected on coastal transects	http://www.ccms.ac.uk/simcoast.htm
Species Analyst	A software extension for ESRI's ArcView GIS software that provide an interface to species distribution prediction models	http://chipotle.nhm.ukans.edu/documentation/applications/SpeciesAnalyst/
SPUR2	A general grassland ecosystem simulation model	HTTP://WWW.GPSR.COLOSTATE.EDU/GPSR/PRODUCTS/SPUR2.HTM
SysTax	Systematic botany and the administration of botanical gardens, herbaria and other collections	http://www.biologie.uni-ulm.de/systax/systax-e.html
TEM	Terrestrial Ecosystem Model	http://baetis.mbl.edu/~dkick/temhp.html
TREEDYN 3	Forest Simulation Model	HTTP://WWW.USF.UNI-KASSEL.DE/DFRAME.PHP3?REFERENCE=/MODELLE/MOD_LEFT.HTM
WORLDMAP	Analysis and mapping	http://www.nhm.ac.uk/science/projects/worldmap/

9

Approaches to spatially distributed hydrological modelling in a GIS environment

L. Olsson and P. Pilesjö[1]

ABSTRACT

Traditionally hydrological models have been based on the drainage basin as the fundamental system delineation and their function have been empirically based. In order to build models capable of describing the movement of water within a drainage basin, a spatially explicit approach is needed (see Chapter 2 for an overview of model types). This paper deals with two aspects of the spatially distributed hydrological model – the atmospheric interface and the geomorphological distribution of water flow. Recent advances in land surface-atmosphere interaction models have improved substantially our ability to estimate the fluxes of water and energy between the atmosphere and the terrestrial ecosystems. One of the most important developments is the use of remotely sensed data for measuring, in a spatially continuous fashion, land surface parameters that can serve as input data to models. Coupling of SVAT (Soil Vegetation Atmosphere Transfer) models with distributed hydrological process (see Chapter 2) models and biological production models can be used to assess the effects of land use/cover changes on the regional hydrological cycle. Recent developments in the processing of digital elevation models for the estimation of flow accumulation and automatic delineation of drainage basins is another important basis for the development of spatially distributed models.

9.1 BASIC HYDROLOGICAL PROCESSES AND MODELLING APPROACHES

Water is the most important limiting factor to vegetation growth, and thereby also one of the most important factors controlling human livelihood. Consequently, modelling within the hydrological cycle has become one of the most important tasks in terrestrial ecology. The aim of the traditional hydrological model has primarily been to predict the amount of discharge from a drainage basin, while water movement within the basin has often been neglected. With the advent of efficient computers and spatial data of high quality, the interest has shifted from those lumped models (see Chapter 2) towards spatially distributed models, where

---

[1] Note that the authors contributed equally to the paper.

water movement within the drainage basin can be modelled. Important applications of this emerging field of spatially distributed hydrological modelling tools include studies of:

- pollution propagation in the soil
- impact of land surface (e.g. agriculture and forestry) management practices on hydrological regimes
- impact of vegetation and land use change on hydrological regimes
- the prediction of nutrient leakage in agricultural landscapes.

The aim of this chapter is to outline the design of distributed models in a GIS environment and to discuss problems and potentials. Another aim is to discuss the use of remote sensing data as an aid in modelling hydrological processes.

The development of a spatially distributed hydrological model can be described as solving three major problems, to be solved in a geographically explicit fashion, these are:

- the partitioning of precipitation into evaporation and water input to the drainage basin
- the partitioning of water input into infiltration and surface runoff
- the movement of surface and subsurface water within the drainage basin.

In the first part of the chapter we describe generally the different components of the hydrological cycle. Understanding of the fundamental processes of water flow is essential in all types of modelling. Even if all models are only generalized mimics of the environment, profound knowledge about the processes helps us to develop and evaluate the models. Apart from the processes, the introductory part also describes different modelling approaches. This section is based on Andersson and Nilsson (1998).

In the second part of the chapter the main input parameters to hydrological models, and mechanisms by which these are derived using GIS and remote sensing, are described. The following section deals with the land surface – atmosphere interface, which describes different model approaches to divide the precipitation into evapotranspiration and water input to the drainage basin. The partitioning of water into infiltration and surface runoff as well as the movement of surface and subsurface water is discussed at the end of the chapter.

9.1.1 The hydrological cycle

There is an unending circulation of water within the environment. This circulation is called the hydrological cycle and its components are presented in Figure 9.1. The energy from the sun evaporates water from open water surfaces and from land. Wind transports the moist air until it condenses into clouds in a cooler environment. Water reaches the ground and water surfaces as precipitation (rain, snow or hail) falling from the clouds. Depending on temperature, the precipitation can be stored as snow, ice or water for a shorter or longer time. A part of the falling rain is captured by the vegetation as intercepted water, which eventually evaporates back to the air. The remainder of the rain will reach the ground or fall into water bodies where evaporation will continue. A portion of the water that reaches the ground may flow directly into streams as overland flow, but most of the surface water will infiltrate into the soil. The water infiltrates until the soil is saturated and cannot hold any more water. If the soil becomes saturated, the excess water will flow on the soil surface as overland flow. The water that has infiltrated into the soil will move downwards or laterally as subsurface flow. The lateral movement is due to diversion when soils of different characteristics are reached or because of differences in water pressure in the water saturated zone (ground water). The downward movement is due to gravity and the water will eventually become part of the ground water. The infiltrating water may also be taken up by vegetation from which it may be transpired back to the atmosphere.

Subsurface, ground water and overland flows contribute to the stream flow which transports the water back to the ocean and completes the hydrological cycle.

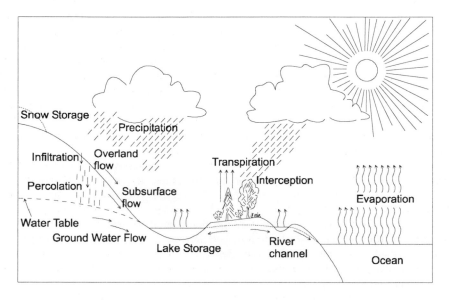

Figure 9.1: Components of the hydrological cycle (from Andersson and Nilsson 1998).

Below follows a short description of the most important components of the hydrological cycle:

Precipitation occurs when moist air is cooled and reaches the dew point temperature. This gives rise to water droplet development on condensation nuclei, e.g. small dust particles.

Precipitation intensity is the amount of precipitation (in liquid form) per time unit.

Interception occurs when vegetation captures precipitation on its path to the ground. The capacity to store water on leaves and stems depends on vegetation type and appearance. The capacity is generally higher for evergreens than for deciduous trees (Selby 1982).

Evaporation is used as a term for the loss of water vapour from water, soil and vegetation surfaces to the atmosphere. The process deals with changes in state of aggregation from liquid water into water vapour and is controlled by the moisture gradient between the surface and the surrounding air. The energy used for the change is primarily net radiation, i.e. from the sun, but can also be taken from stored heat in e.g. vegetation or water bodies.

The water vapour capacity of the air is directly related to temperature (Shaw 1993).

Evaporation from soil originates from temporary surface puddles or from soil layers near the surface. The effectiveness of the evaporation depends on the aerodynamic resistance, which in turn is dependent upon wind speed, surface roughness, and atmospheric stability, all of which contribute to the level of wind turbulence (Oke 1995).

Transpiration is the water loss from the soil through the vegetation. Transpiration differs from evaporation since vegetation can control its loss of water. Plants draw their water supply from the soil, where the moisture is held under pressure. They control the rate of transpiration through the stomata in their leaves by changing the area of pore openings. Usually this factor is referred to as stomata resistance, and depends on the water content of the air, the ambient temperature, the water availability at root level, light conditions, and carbon dioxide concentration (Oke 1995). The pores close in darkness and hence transpiration ceases at night. When there is a shortage of water in the soil the stomata regulates the pores and reduces transpiration. Transpiration is thus controlled by soil moisture content and the capacity of the plant to transpire, which in turn are conditioned by meteorological factors (Shaw 1993).

If there is a continuous supply and the rate of evaporation is unaffected by lack of water, then both evaporation and transpiration are regulated by the meteorological variables radiation, temperature, vapour pressure and wind speed (Shaw 1993). When the vegetation is wet the loss of water is dominantly due to evaporation. During dry conditions the water loss from vegetation surfaces is mainly via transpiration. Some 20–30 per cent of the evaporated water originates from intercepted vegetation storage when such occur (Lindström *et al.* 1996).

Usually the combined loss of water from ground, water surfaces and vegetation to the atmosphere is called *evapotranspiration*.

Infiltration is generally described as the penetration and flow of water into the soil. When a soil is below field capacity, which is the capacity of water content of the soil after the saturated soil has drained under gravity to equilibrium, and precipitation is gathered on the surface, the water penetrates into the soil. The water infiltrates at an initial rate dependent on the actual soil moisture content and the texture and structure of the soil. As the precipitation supply continues the rate of infiltration decreases, as the soil becomes wetter and less able to take up water. The typical curve of infiltration rate with time reduces to a constant value, called the infiltration capacity (Shaw 1993), which usually is equal to, or slightly less than, the saturated hydraulic conductivity. The hydraulic conductivity is a measure of the water leading capability of the soil, and is controlled by the soil pore size, soil composition, and the soil moisture content. The saturated hydraulic conductivity is often referred to as permeability (Grip and Rhode 1994).

The actual infiltration capacity of a soil varies depending on the soil characteristics and the soil moisture. Pre-existing soil moisture is an important infiltration regulating factor because some soils exhibit an initial resistance to wetting (Selby 1982).

During infiltration the soil is getting soaked, but when the rainfall stops soil beneath the wetting front is still getting wetter while soil above is drying as it drains. The type of vegetation cover is also an important factor influencing infiltration. Denser vegetation results in higher organic concentration in the root zone, promoting a thicker soil cover and a more loose structure. This results in a higher infiltration capacity. Vegetation and litter also decrease the precipitation impact on the surface. Without these factors smaller particles would be thrown into suspension, and clogging might occur as they are re-deposited and less permeable layers would evolve (Chorley 1977).

Overland flow is often divided into Hortonian overland flow and saturated overland flow. When precipitation intensity exceeds the infiltration capacity, the precipitation still falling on the area can not infiltrate and the excess water flows on the surface. This kind of flow is referred to as Hortonian overland flow. The second type of overland flow occurs when completely saturated soils give rise to saturated overland flow without having precipitation falling upon it, due to pressure effects induced by subsurface water infiltrated in up-slope areas. The velocity of overland flow depends on the slope angle and the surface roughness.

Subsurface flow occurs below the soil surface. Close to the surface, where the soil normally is not saturated, we have an unsaturated water flow. At greater depths the soil reaches saturation (i.e. the water pressure exceeds the atmospheric pressure). The surface at which the pressure equals the atmospheric pressure is defined as the groundwater table (see Figure 9.2). Below the groundwater table all soil pores are completely filled with water. This is referred to as the saturated zone. The connected pore system in a soil can be seen as small pipe shaped areas and the water level rise is referred to as the capillary rise. The zone is often called the capillary fringe (Grip and Rhode 1994). The extent of the capillary fringe is dependent on the soil composition and the packing of the soil particles. It ranges from a few centimetres in a coarse sandy soil to several meters in a clay soil. The soil above the capillary fringe is referred to as the unsaturated zone or the aeration zone and has pores filled with a mixture of water, water vapour and air. After e.g. heavy rains, parts of the unsaturated zone might become temporarily saturated.

As shown in Figure 9.2 unsaturated flow occurs in the unsaturated zone. The water with a vertical flow inside the unsaturated soil is usually referred to as *percolation*. The velocity of the diverted flow is dependent on soil permeability and stratum slope (Brady and Weil 1996).

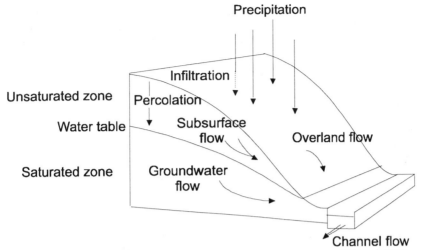

Figure 9.2: Overland and subsurface flow (from Andersson and Nilsson 1998).

In an unstratified soil there is also a tendency for more compaction and smaller pores with greater depth. This leads to successively lesser permeability and a greater partition of the water will be forced to move sideways as subsurface flow (Chorley 1977). The diverted flow is saturated, but not included in the saturated zone since there is unsaturated soil beneath it (Andersson and Burt 1985).

When the percolating water reaches the saturated zone, or more exactly the capillary fringe, the water is incorporated with the groundwater (see Figure 9.2) and the water table is temporarily raised. Inside the saturated zone spatially different water pressures govern the water movements. The theory for groundwater movement is based on Darcy's law (see below). An important extension of Darcy's law for groundwater flow is its application in three dimensions. The permeable material, the soil, is often heterogeneous. Clay layers can for example be present in sandy material, and soil close to the surface is often more porous than material at greater depth. The hydraulic properties of the ground are not isotropic, and the hydraulic conductivity is different in different directions (Bengtsson 1997). The discharge rate is therefore into three perpendicular discharges (Q_x, Q_y and Q_z) with different hydraulic conductivity and different flow velocities (Shaw 1993).

The groundwater flow is always in motion but at very slow velocities, about three magnitudes lower than overland flow (Chorley 1977). At a large scale the groundwater movement is directed from a recharge area to a discharge area. Recharge areas can be defined as areas with a vertical flow component downwards inside the groundwater zone. Discharge areas are defined as the opposite phenomena, where the principal water flow is upwards.

Once water has infiltrated into the ground, its downward movement to the groundwater and the amount of stored groundwater depends on the geological structure as well as on the rock composition. In general, older rock formations are more consolidated and the rock material is less likely to contain water. Igneous and metamorphic rocks are not good sources of groundwater, unless weathered and fractured. The sedimentary rock strata have different composition and porosity and are much more likely to contain large amounts of water (Shaw 1993). Beds of rock with high porosity that are capable of holding large quantities of water are often referred to as aquifers. Aquitards are semi-porous beds, which allow some seepage of water through them. Clay beds, which are almost impermeable, are called aquicludes (Shaw 1993).

9.1.2 Modelling approaches

The two classical types of hydrological models are the deterministic and the stochastic, where stochastic models involve random elements (see also Chapter 2). The deterministic models can be classified according to whether the model gives a lumped or a distributed description of the considered study area. The models can also be classified whether the description of the hydrological processes is empirical or physically-based. There are three major types of deterministic models: empirical lumped models, empirical distributed models, and physically-based distributed models. The fourth combinatory possible type would be the physical lumped model, but this concept is somewhat contradictory since physical models require measurable input data whereas lumped models use averages for an entire catchment. Classifications of this kind are however fluent. Lumped models might have parameters that are more or less distributed. Models can also have components with both physical and empirical origins, so called semi-empirical or grey box models (Abbott and Refsgaard 1996).

Empirical models (see section 2.4.1) are based on regression and correlation results from statistical analyses of time series data. The derived equations are based on observed phenomena or measurement knowledge without demands on understanding of the underlying processes. Empirical models are often referred to as black box models. Truly *physical models* (see section 2.4.3) are based on formulas of physical relations. They are analogously referred to as white box models (Kirby *et al.* 1993) since every part of the processes is understood. The input data include only measurable variables that can be collaborated.

Physically-based models are the most suitable when studying internal catchment change scenarios. Examples of this are irrigation and groundwater use development. The prediction of discharge from catchments including monitoring of pollutants and sediments dispersed by water are also well suited for physical models (Andersson and Burt 1985; Abbot and Refsgaard 1996). It is important to note that not all of the conceptual understanding of the way hydrological systems work is expressible in formal mathematical terms. Thus any model definition will be an abstraction of the total knowledge of catchment hydrology. Thereby all models include a systematic error based on the not included or not known relationship. This is a neglected source of error in many physical modelling processes, and yields a need of calibrating the model to time series data. Practically, this means we have very few truly physically-based models but many semi-physical ones.

A *lumped model* (see section (2.4.3) operates with interrelated reservoirs representing physical elements in a catchment being the smallest spatial element in the modelling system. This results in that the model uses parameters and variables that represent average values for the entire catchment. These averages can be derived either physically or empirically which can give the model a semi-empirical appearance. Lumped models are mainly used in rainfall-runoff modelling.

Distributed hydrological models (see section (2.4.3) are supposed to describe flow processes in each and every point inside a catchment. Due to difficulties within the general conceptual modelling framework and very time and memory consuming programs these models are practically impossible to use. Simpler models instead try to estimate the different flow patterns discretised into nodes with orthographic spacing. These nodes can be seen as centre points in square shaped areas referred to as pixels or cells. If a model is based on this type of cell structure it is directly compatible with remotely sensed and gridded (raster) GIS data. In the vertical extent each orthographic cell might be given a depth, or be discretised into a number of overlaying cells (i.e. a column). For each cell the water discharge to neighbouring cells is calculated according to the active hydrological processes. The flow distribution inside the catchment is thereby mapped. Even if the processes are estimated as a continuum, the stored results are discretised into cells (Abbott and Refsgaard 1996).

The distributed nature of a modelling system means that spatial variation, characteristics and changes can be simulated and estimated inside a catchment. Distributed hydrological models have particular advantages in the study of the effects of land use changes. The model not only provides a single outlet discharge, but multiple outputs on a temporally and spatially distributed basis. The disadvantages with this form of modelling are the large amounts of data and the heavy computational requirements. The model type also includes a large number of parameters and variables, which have to be evaluated. The effect of scale choice (cell size) is also an uncertainty (Beven and Moore 1993).

A *stochastic model* (see section 2.5) uses random elements, which are drawn from statistically possible distributions. This means that the simulations will not give the same results when repeated with the same input data. With most stochastic models the approach is to conduct a multitude of simulations, the so-called Monte Carlo technique, and produce average estimates with specified confidence intervals.

9.2 DATA FOR SPATIALLY DISTRIBUTED HYDROLOGICAL MODELLING

One of the most severe problems to overcome in distributed hydrological modelling is the mismatch of scales between processes and obtainable data, both in terms of spatial scales as well as temporal scales. The most important divide in relation to data sources is between point data and spatially continuous data. Most of the climatic data necessary for hydrological modelling can only be obtained at a point basis, even though remote sensing methods are becoming increasingly important. But on the other hand, the point data are often available at very short time intervals (hours). Data on subsurface properties, soil and rock conditions, are also primarily point based and subsequently extrapolated to cover a region. Concerning vegetation, topography and surface conditions, spatial continuously data are often available, but with varying resolution in time and space. Data on topography necessary for the studies of water movement within a catchment are typically available at a spatial resolution of 25 m to 50 m, which corresponds well with data on vegetation and surface conditions available from high-resolution remote sensing (e.g. Landsat and Spot). However, the temporal resolution of these remote sensing data (typically yearly, considering costs and other practical factors) are too coarse to capture the biological aspects of the hydrological cycle. Temporal resolution adequate for studying vegetation and climatic processes (from bihourly to bimonthly) are available, but at a much coarser resolution, typically 1 to 5 km. In order to make the best out of these conflicting scales, in time and space, profound knowledge on data sources and handling coupled with a large portion of creativity are needed.

9.2.1 Vegetation

In order to successfully set up and run a distributed hydrological model the vegetation must be described by appropriate parameters in a spatially explicit fashion. Over large regions, the only practical means is by remote sensing. The vegetation parameters needed relate primarily to the role of vegetation in the following processes:

- evaporation and transpiration
- interception
- infiltration .

Here we will concentrate on the first two processes.

We can distinguish between two approaches to estimate vegetation parameters from remote sensing (see also Chapter 6):

- to infer vegetation parameters directly from the remote sensing data, or
- to classify vegetation types and model vegetation parameters independently of the remote sensing data.

The first approach requires time series of remote sensing data throughout the vegetation season that is usually only available at a coarse spatial resolution. The most important remote sensing data sources for time series data are the NOAA AVHRR sensing system and different geostationary satellites, e.g. Meteosat covering Europe and Africa. Only the vegetation parameters inferred directly from remote sensing data will be discussed in this chapter.

Satellite sensors provide us with a continuous flow of data on the amount of reflected and emitted radiative energy from the Earth. For studies of vegetation dynamics, the most important part of the spectrum is in the visible and the near infrared region, with special focus on the photosynthetically active radiation (0.4 – 0.7 μm), usually referred to as PAR. Figure 9.3 shows the most important flows of PAR that are used in remote sensing.

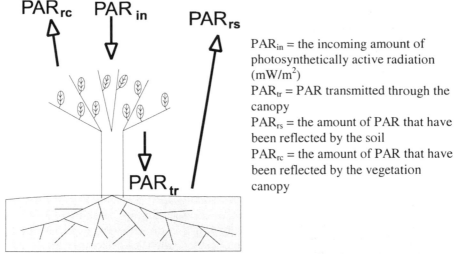

PAR_{in} = the incoming amount of photosynthetically active radiation (mW/m^2)
PAR_{tr} = PAR transmitted through the canopy
PAR_{rs} = the amount of PAR that have been reflected by the soil
PAR_{rc} = the amount of PAR that have been reflected by the vegetation canopy

Figure 9.3: The partitioning of incoming photosynthetically active radiation.

We can then define the important parameter absorbed PAR, APAR, according to:

$$APAR = \left(PAR_{in} + PAR_{rs}\right) - \left(PAR_{rc} + PAR_{tr}\right) \qquad (9.1)$$

Green plants use water (from the roots), carbon dioxide (from the atmosphere) and energy (from the sun) as input to the photosynthesis, and if we can determine the amount of energy the plants are using, we have a link to measure the rate of photosynthesis, which is also an important link to other biophysical processes related to hydrology.

A fundamental problem in remote sensing is to distinguish between the vegetation fraction of the signal from the soil fraction of the signal. This is usually approached through the construction of a vegetation index, where we can distinguish two principally different kinds of indices, ratio-based indices and orthogonal-based indices (for a comprehensive discussion of different vegetation indices, see Huete (1989) and Begue (1993)). The normalized difference vegetation index, NDVI, is the one that has become the most widely used index, and it is defined as equation 9.2.

$$NDVI = \frac{\left(NIR - RED\right)}{\left(NIR + RED\right)} \qquad (9.2)$$

where *NIR* and *RED* is the amount of reflected light of visible red and near infrared wavelengths respectively.

A number of studies, from ground based measurements as well as from satellite based ones, have confirmed a linear relationship between NDVI and the fraction of absorbed PAR to the incoming PAR, a parameter usually denoted fAPAR and defined as in equation 9.3.

$$fAPAR = \frac{APAR}{PAR_{in}} \qquad (9.3)$$

Some empirical relationships between fAPAR an NDVI presented in the literature are shown below.

fAPAR = 1.42 *   NDVI –   0.39   (Lind   and   Fensholt   1999)
fAPAR = 1.41 *   NDVI –   0.40,   r2   =   0.963   (Pinter   1992)
fAPAR = 1.67 *   NDVI –   0.08   (Prince   and   Goward   1995)
fAPAR = 1.62 *   NDVI –   0.04, r2 = 0.96 (Lind and Fensholt 1999)

Regression slopes found in the literature usually are between 1.2 and 1.6 with intercepts between -0.02 and -0.4. The linear relationship between NDVI and fAPAR have been found to be remarkably consistent over a range of non-woody vegetation types but the relationship deteriorates when LAI becomes more than about 2.

The derivation of leaf area index is less straightforward than the fAPAR transformation above, and the best result is probably obtained by the use of a full radiative transfer model, see for example: Begue (1993) and Myneni and Williams (1994), but for operational use an approach based on Beer's law can be used. Beer's law expresses the relationship between the amount of PAR transmitted through a vegetation canopy and the LAI, according to:

$$PAR_{tr} = e^{\frac{-G \cdot LAI}{\sin b}} \qquad (9.4)$$

where,

b = solar elevation (could be replaced by solar zenith angle, θ (sin b = cos θ))
G = mean direction cosine between solar zenith angle and leaf normals.

A relationship between LAI and fAPAR can then be established (Sellers *et al.* 1996) which scales the LAI logarithmically as a function of fAPAR. The maximum LAI for a particular vegetation type is set to the corresponding maximum fAPAR for that vegetation, according to the formula below.

$$LAI = LAI_{max} \cdot \frac{\log(1 - fAPAR)}{\log(1 - fAPAR_{max})} \qquad (9.5)$$

where,

LAI_{max} = the maximum possible LAI for a particular vegetation type
$fAPAR_{max}$ = the corresponding fAPAR for the LAI_{max}.

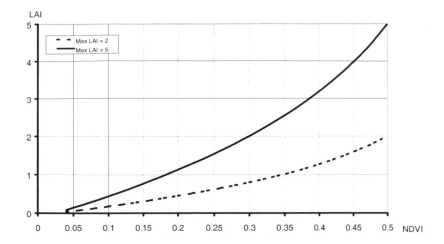

Figure 9.4: Leaf area index as a function of NDVI for two vegetation types with maximum LAI of 2 and 5 and with a corresponding maximum fAPAR of 0.77.

Typical ranges of LAI found using this relationship over the Sahel region in Africa using NOAA AVHRR data are shown in Table 9.1.

Table 9.1: Typical LAI values found in the Sahel region of Northern Africa by transforming NDVI to fAPAR and fAPAR to LAI using equation 9.5. Data from NOAA AVHRR.

Vegetation class	Minimum LAI	Maximum LAI
Evergreen broadleaf forest	0.6	3
Closed shrublands	0.1	0.7
Open shrublands	0	0.15
Woody savannahs	0.4	2
Savannahs	0.2	1.8
Grasslands	0	0.5
Croplands	0	1.7
Cropland and natural vegetation mosaic	0.1	1
Barren or sparsely vegetated	0	0.1

Leaf area index plays an important role in many processes related to how vegetation controls the movement of water from the roots, through the plant and from the plant to the atmosphere. It is also important in determining the amount of water intercepted by a vegetation canopy.

Interception of water by the vegetation canopy is a very complex process to estimate and different approaches to interception estimation have been tried, for a discussion of these, see Gash (1979), Calder (1986) and Jones (1997). One very simple approach has been taken by the SHE model (Abbot and Refsgaard 1996), and is expressed as,

$$I_{max} = C_{int} \cdot LAI \qquad (9.6)$$

where
I_{max} = interception storage capacity
C_{int} = empirical interception parameter, typically 0.05 mm for deciduous forest.

9.2.2 Modelling vegetation growth

It is obvious that a realistic modelling of water balance needs a coupling of water balance models and vegetation growth models. The most promising approach is to use a top-down modelling driven by a combination of remote sensing data and ground observations.

On of the most widely used methods for quantification of vegetation growth is the one that relates net primary productivity (NPP) to the accumulated amount of absorbed photosynthetically active radiation (APAR) over one year or one vegetation season.

$$NPP = \sum_{i=1}^{365} \varepsilon_i \cdot APAR_i \quad (9.7)$$

where
ε_i = a factor accounting for photosynthetic efficiency
$APAR_i$ = absorbed photosynthetic active radiation.

For a remote sensing based solution we may use the previously described relationship between fAPAR and NDVI (equation 9.3).

Combining the equations yields the equation:

$$NPP = \sum_{i=1}^{365} \left[\varepsilon_i \cdot (a \cdot NDVI_i + b) \cdot PAR_{in_i} \right] \qquad (9.8)$$

where PAR_{in} = the incoming amount of PAR, which is a function of latitude, time and cloudiness.

The real challenge is then to find ways of estimating the efficiency factor, which is a factor varying with vegetation stress due to water, temperature and nutrient deficiencies. We can distinguish two different approaches to estimate the stress factor, remote sensing based methods involving the use of thermal and visible/NIR sensors (Asrar 1989), and indirect estimation through biophysical models (for example like CENTURY, CASA, SiB, BATS etc.) (Sellers 1992; Prince and Goward 1995; Lind and Fensholt 1999).

9.2.3 Topography

The first step in hydrological modelling is to define a model area by delineating the outline of the catchment boundary. Since our modelling approach is distributed, and more topographic information is needed (e.g. slope and drainage area in order to estimate overland flow, see below) we normally use a digital elevation model (DEM) for the estimations of topographically related parameters.

A DEM is a matrix where every cell value represents the elevation at the centre point in the corresponding area on the Earth's surface. Normally the DEM is interpolated from line or point data (e.g. paper maps or point measurements), but can also be constructed by the use of digital photogrammetry. The quality of the DEM is crucial for the estimation of the topographic parameters as well as for the reliability of the model output.

The quality of the DEM depends on numerous things, but three major factors can be easily distinguished:

- the used interpolation algorithm
- the spatial distribution of the input data points
- the quality (x, y, z) of the input data points.

We can interpolate any spatially distributed variable. After the interpolation, a continuous surface in three dimensions is created. The third dimension of the surface is the value of the measured variable (in our case elevation).

Interpolation algorithms can be divided into two types: global and local methods. The global interpolation methods use all in-data points when estimating the surface, and the results show general trends in the data material. These algorithms are normally not suitable for interpolation of topographical data. Preferably we use a local method, which only use a limited number of in-data points, close to the cell that we are interpolating, when working with topographical data. The obvious reason to use a local method is their sensitivity to smaller, locally distributed terrain forms that are often present in natural terrain. Examples of local interpolation methods are Thiessen polygons, inverse distance interpolation, spline functions and kriging (see Burrough and McDonnell 1998).

The choice of interpolation algorithm primarily depends upon the spatial autocorrelation of the in-data points, and different methods can produce significantly different results. However, the use of geo-statistical interpolation (e.g. kriging), where the user examine the spatial variation in the data and let the autocorrelation in different directions guide the interpolation, is often recommended for interpolation of evenly distributed point data.

As mention above, the distribution of the input data points is extremely important for the result of the interpolation, and consequently for the accuracy of the DEM. Most interpolation algorithms are very sensitive to the distribution of in-data, and demand evenly distributed data to produce a reliable result. Even if some algorithms are more sensitive than others, the importance cannot be stressed enough.

One very useful approach to handle data unreliability is to use Monte Carlo simulation (i.e. a stochastic model – see Chapter 2) when generating a DEM. This is a method which is conceptually very simple, but very computer intensive. The principle is to repeat a calculation many times, and each time a random element is added to the input data. The method is particularly well suited for applications where it is impossible to analytically calculate the error in every data point, but where it is possible to estimate the overall variability of the data set. One such case is the task to delineate watersheds using DEMs. If we can estimate the overall degree of confidence in the elevation data used for interpolation of a DEM, expressed as an RMS error, we can then use this RMS estimation to generate many elevation models, each with a random element added to it. If we carry out the drainage basin delineation on each DEM, we will get the final result as the sum of all the delineations. Using this method we obtain not just a more reliable drainage basin delineation, but we also get an idea of the confidence of the result. One such example is shown in Figure 9.5, where 100 DEMs have been interpolated from topographic data with 5 m contour intervals. The variability of the data was then estimated to a normal distribution with a standard deviation of ± 5 m. Each of the drainage basin delineations resulted in a raster file coded with 0 for outside and 1 for inside the basin. The sum of all delineation results yields a file with values between 0 and 100, where 0 means outside the basin in all DEMs and 100 means inside the basin in all DEMs.

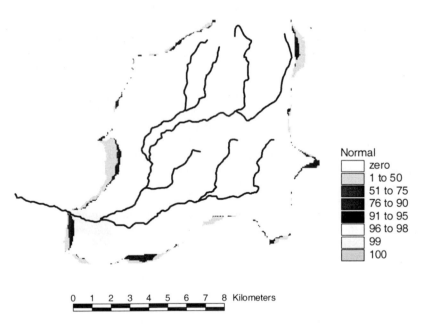

Figure 9.5: Drainage basin delineation based on Monte Carlo simulation. The figure shows the result of 100 basin delineations.

9.2.4 Soil

The important soil characteristics needed for hydrological modelling relate primarily to two aspects of the hydrological processes,

- factors governing the water holding capacity and the movement of water within the soil, and
- factors governing the infiltration capacity and infiltration velocity.

The latter factors have been proven to be of particular importance in the dry tropics, where crusting and sealing of the soil surface can have a large effect on the amount of water actually passing into the soil volume. In hot and dry climates, a slowing down of the infiltration during a rain event will increase the evaporative losses tremendously. The infiltration characteristics of a soil are often very hard to estimate, and fairly rough parameterisation schemes are often used. It is important to point out that infiltration is to a high degree altered by land management and vegetation growth and therefore varies with time. Some examples of infiltration capacities are shown in Table 9.2.

Table 9.2: The effect of different agricultural practices on infiltration capacity (mm/h) (modified from Jones 1997).

Soil type / management	Management / time	Infiltration capacity (mm/h)	Management / time	Infiltration capacity (mm/h)
Silt loam, high organic content:	Under pasture:	27.3	Under cornfield:	6.8
Silt loam, low organic content:	Under pasture:	8.3	Under cornfield:	6.5
Cornfield, unmulched:	In June:	22	In October	6.0
Cornfield, mulched:	In June:	42	In October	39
Coarse soil on sandstone:	Grazed rangeland:	32	Ungrazed:	48
Fine soil on shale:	Grazed rangeland:	18	Ungrazed:	17
	Unimproved pasture:	43	Oak-hickory forest:	76

The factors governing the amount of water that can be held, and the movement of water in the soil are of course difficult to obtain data on, and we have to rely on a limited number of soil properties from which the desired characteristics might be inferred. Soil texture classification, often in the form of fractions of clay, silt and sand, is the single most important data source, and it is also the most commonly available type of soils data. From the texture classification we usually infer estimates of hydraulic conductivity, matric potential and saturated water content. Depth to the underlying rock or other impermeable layer is often desirable, but difficult to obtain.

9.2.5 Climate

A range of different climate variables is needed, and the more advanced model the more demand is put on the climatic data input. For a model to be applied in different environments it is important to restrict the climate data requirements to what is generally available from operational climate stations. But the role of remote sensing for measuring or inferring climate variables is increasing.

The time interval of data input is also a matter for concern. Many detailed models require daily data, but it would often be desirable to use bi-daily data in order to capture the important diurnal variability. A list of common climate variables for a comprehensive modelling of evaporation using the Penman-Monteith formula (see further below) could look like as shown in Table 9.3.

Table 9.3: Example of climate variables needed for evaporation calculations using a Penman-Monteith type of model.

Climate variable	Unit	Source of data
Average daytime temperature	K	observations, max, min and length of day
Cloud cover	%	nearest station or satellite data
Dew-point temperature	K	observations, or min temp. approximation
Average daily wind speed	m/s	nearest station
Precipitation	Mm	observations (snow is inferred from temp.)
Precipitation intensity	mm/h	observations, but often subjective approximations
Relative humidity	%	Observations

Remote sensing based methods for estimating the surface energy balance will play an important role in the near future for our possibility to model climate parameters between observation points. The energy balance plays a very important role for the estimation of evaporation, and substantial progress have been made during the last 10 years to model actual evaporation rates by means of thermal remote sensing. For a detailed discussion of methods, see Kustas *et al.* (1989).

Net radiation can be expressed as the sum of four major components, i.e. the downward and the upward short- and longwave radiation components,

$$R_{net} = Rs_\downarrow - Rs_\uparrow + Rl_\downarrow - Rl_\uparrow$$

The two downward fluxes are usually estimated from point observations, which can be extended over relatively large areas (100 km² for shortwave and 10 km² for longwave under stable weather conditions (Kustas *et al.* 1989), while the two upward components can be inferred by varying success from remote sensing data.

The shortwave reflected radiation, the albedo, can be estimated by means of remote sensing methods (Otterman and Fraser 1976; Wanner *et al.* 1995). Two problems, though, are associated with remote sensing methods. Firstly, remote sensing systems (e.g. Landsat TM or NOAA AVHRR) only measures about 50 per cent of the solar spectrum, and secondly albedo is often quite variable in different angles, whereas most remote sensing methods only measure in one angle. The emitted longwave radiation can be estimated remotely by means of sensing systems in the thermal IR region, typically 10–12 µm. The approach was successfully demonstrated by Jackson (1985), but more research is needed within this field.

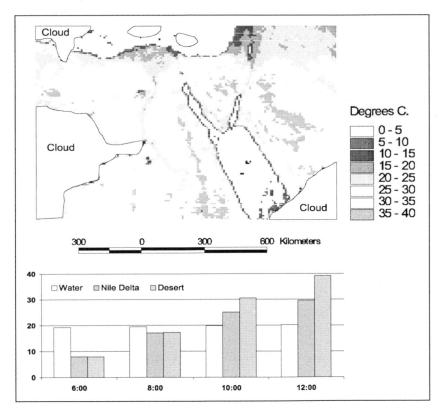

Figure 9.6: The Nile valley in Egypt seen from Meteosat on 1 May 1992. The image shows temperature change (degrees C) from 6 am to 12 noon.

An important concept in modelling evaporation, particularly in dry and hot climates, is thermal inertia, which can be measured by means of temperature change over time. Satellite systems available for thermal inertia studies are e.g. NOAA AVHRR sensor for diurnal changes and the geostationary weather satellites (like Meteosat) for almost continuous temperature measurements. The diurnal temperature change of the land surface is to a large extent governed by the evaporation of water, particularly in arid and semi-arid environments (insolation effects not considered), and by continuously measuring the diurnal temperature variation (a function of the thermal inertia) there is a good hope of being able to estimate the actual rate of evaporation. More research as well as improved satellite sensors are needed. One example of the temperature change during the morning hours, showing the effect of moisture on the thermal satellite data is shown in Figure 9.6.

In Figure 9.6 it is obvious that the moisture availability along the Nile valley has a cooling effect on the temperature compared with the surrounding desert areas.

9.3   THE LAND SURFACE – ATMOSPHERE INTERFACE

9.3.1 Estimation of potential evaporation

Potential evaporation, i.e. the evaporation from an open water surface, is a fundamental process in the hydrological cycle. A number of different approaches to partition precipitation into evaporation and input to the terrestrial ecosystem have been employed over the last centuries. In the early nineteenth century, John Dalton formulated the law that bears his name and states that evaporation, E, can be expressed as:

$$E = u \cdot \left(e_s - e_a \right) \qquad (9.9)$$

where
u = function of wind speed
e_a = the current vapour pressure
e_s = the saturation vapour pressure at that temperature.

Dalton assumed that over water or a moist surface, the air immediately above it rapidly becomes saturated and prevents further evaporation. That is why the wind speed is important. This approach really only describes the potential evaporation, i.e. free evaporation from a constantly wet surface. The actual evaporation on land depends on a range of factors: the depth to the water table, the porosity of the soil and the local heat budget. Transpiration is controlled by a number of additional factors related to the vegetation. From a hydrological point of view, we often add the evaporation and the transpiration to one variable, called evapotranspiration, ET.

Since the formulation of Dalton's law in 1802, a great number of formulae for the calculation of evaporation have been developed. What determines the type of evaporation formula to apply depends to a large extent on the data availability.

The simplest methods are the temperature based ones, where the data needed usually are restricted to temperature, and in some cases sunshine hours. The most widely used formula of this kind is by Thornthwaite (Thornthwaite and Mather 1955), expressing monthly evaporation (cm) as:

$$E = 1.6 \cdot d \cdot \left(\frac{10 \cdot \bar{t}}{I} \right)^a \qquad (9.10)$$

where
d = total monthly daylight hours/360
t = mean monthly temperature
$I = \Sigma i$ where $i = t^{1.514}$
a = empirical coefficient.

Another approach, more physically based and which corresponds closely to Dalton's law, is the mass transfer approach, where the data requirements also include wind speed and some measure of air humidity. The energy budget approach is based on the assumption that all surplus energy from the net heat balance is used for evaporation. The most sophisticated formulae for calculation of potential evapotranspiration are the combined formulae, which are combinations of the mass transfer and the energy balance approaches. The Penman formula from 1948 (Penman 1948) has been one of the most widely used methods.

The Penman formula considers all the important components of the evaporation process from a soil or water surface, but keeps the data requirement to readily accessible meteorological variables. The formula has been widely used all over the world, but it has also been modified to make the best use of available data. The basic form of the Penman formula is:

$$E = \frac{\Delta \cdot R}{\Delta + \gamma \cdot \rho \cdot L} + \frac{\gamma}{\Delta + \gamma} \cdot f(W) \cdot (1 - RH) \cdot e_a \qquad (9.11)$$

where
Δ = rate of change of saturated water vapour pressure with temperature
γ = psychromatic constant (fairly constant around 0.6)
ρ = density of water
R_n = net radiation
RH = relative humidity
e_a = actual water vapour pressure.

The wind function f(W) is calculated according to the formula below:

$$f(W) = a \cdot (1 + b \cdot W) \qquad (9.12)$$

where
a = constant often set to 0.13 (mm/day, mb)
b = constant often set to 1.1 (s/m) over land and 0.5 over water (Kuzmin 1961)
W = wind speed (m/s).

The first term in the Penman equation is the energy balance, in which the energy available for evaporation ignores heat storage. The net radiation, R is usually calculated from the energy balance equation according to:

$$R = (1-\alpha) \cdot R_s + (1-\alpha_l) \cdot R_{l\_in} - R_{l\_out} \qquad (9.13)$$

where
α = albedo (subscript l means longwave)
R_s = incoming shortwave radiation
R_{l_in} = incoming longwave radiation
R_{l_out} = outgoing longwave radiation .

A further refinement of the Penman equation is the Penman-Monteith formula (Monteith and Unsworth 1990), in which the role of vegetation has been further elaborated. The extra variables taken into consideration are aerodynamic resistance, and is considered in the f(W) term, and the stomata resistance which controls the diffusion of water through the leaf. These variables are reasonably well known for agricultural crops, but for natural ecosystems it is difficult to find appropriate values.

The Penman-Monteith equation has been used and modified by many scholars. One form of the equation is presented here (Monteith 1981):

$$\lambda E = \frac{1}{\Delta + \dfrac{\gamma \cdot (r_{sfc} + r_{aero})}{r_{aero}}} \cdot \left[(R_n - G) \cdot \Delta + \frac{\rho \cdot c_p \cdot (e_s - e)}{r_{aero}} \right] \qquad (9.14)$$

where
λE = the average daytime latent heat flux (evapotranspiration) (W/m^2)
λ = latent heat of vaporization (J/kg)
E = moisture mass flux (kg/s)
γ = psychromatic constant (Pa/K)
(e_s-e) = the average daytime vapour pressure deficit between a saturated surface at the temperature of the air and the ambient vapour pressure (Pa)
ρ = the atmospheric density (kg/m^3)
c_p = specific heat of water (J/kg, K)
r_{sfc} = surface resistance to evaporation (s/m)
r_{aero} = atmospheric resistance to evaporation, (s/m)
$(R_n - G)$ = net radiation minus ground heat flux (W/m^2).

The role of vegetation is introduced in the calculation of the atmospheric and surface resistances to evaporation, which are functions of the vegetation's leaf area index, root distribution and height. To calculate the atmospheric resistance we need to have information on wind speed and vegetation canopy height, while the surface resistance also needs information on leaf area index. A number of different formulae have been suggested to relate vegetation characteristics (height and LAI) to the surface and aerodynamic resistances, and the reader is referred to Monteith and Unsworth (1990) and Guyot (1998) for a comprehensive discussion on this issue. Some guidelines for the values of the canopy resistances are the following: for wet vegetation, $r_{sfc} = 0$ (i.e. the Penman-Monteith Equation is equal to the Penman Equation), for a ley crop, r_{sfc} can be between 10 and 30 (s/m) (Bengtsson 1997) and for a pine forest Lindroth (1984) found a value of $r_{sfc} = 100$ as appropriate.

9.3.2 Estimation of actual evaporation and evapotranspiration

In order to estimate potential evapotranspiration as well as actual evapotranspiration, we need either direct flux measurements or it may be possible to infer this from remote sensing measurements. It is also possible to modify the potential evapotranspiration models above by a factor describing the moisture availability. A number of approaches have been taken here.

The first the most simple one is the so called bucket model (Budyko 1969), in which the terrestrial ecosystem was treated as a bucket with a maximum capacity equal to field capacity, FC. The bucket fills when precipitation exceeds evapotranspiration and surface runoff happens when the water level exceeds the field capacity. The actual evaporation is then estimated as a function of the amount of water in the bucket. When the soil water content reach wilting point, WP, the evapotranspiration also becomes zero.

$$Et = m \cdot \lambda E \qquad (9.15)$$

where
Et = actual evapotranspiration
m = moisture availability.

Scaling of the m parameter is mainly a function of the water holding properties of the soil and one simple scaling scheme (Bergström 1992) is shown in Figure 9.7.

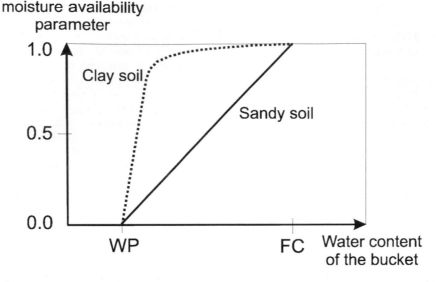

Figure 9.7: The scaling of the moisture availability parameter as a function of soil water content for two different soil types.

The bucket model can be made quite realistic if the evapotranspiration function considers both bare soil evaporation and transpiration through a vegetation canopy, but the real weakness is that the vegetation is still treated as a static component.

The second generation of these models is often referred to as Soil-Vegetation-Atmosphere-Transfer models (SVAT), and the main difference from the first generation models is that vegetation is modelled as dynamic entity. The important vegetation parameters needed for this kind of models is stomata/canopy conductance. Conductance is usually inferred from the absorption of PAR, i.e. photosynthesis rate, which makes sense since water escapes the leaves during opening of stomata. This type of model is sometimes referred to Green SVAT models and for a comprehensive treatment see Sellers (1992). The weakness of these models is that there is no spatial interaction between neighbouring areas, and this is where a spatially distributed hydrological model comes in. The development of green SVAT models coupled to a spatially distributed model for the movement of water on and under the surface, which is still very much at the research stage, is sometimes referred to the 'wetting' of SVAT models.

SVAT models of different degrees of complexity are used in climate models, where there is a coupling between the climatic components of the model and the terrestrial ecosystem through the SVAT model. SVAT models are not yet in operational use in hydrological models, but we anticipate a rapid development within this field.

9.4 THE DISTRIBUTION OF SURFACE FLOW IN DIGITAL ELEVATION MODELS

The amount of water transported as Hortonian overland flow depends, apart from the difference between the precipitation intensity and infiltration capacity, on the potential storage on the ground (Bengtsson 1997). The storage capacity is equal to the volume of small depressions, often called sinks or pits, creating isolated small drainage basins over the surface. The volumes of these sinks can often be estimated by the use of a standard GIS.

The velocity of the overland flow depends on the slope of the surface, the surface roughness (symbolized by a friction factor, f), and the water depth (Bengtsson 1997). The relationship between flow velocity, surface roughness, flow depth and slope can be expressed by Manning's equation:

$$v = M * R^{2/3} * I^{1/2} \qquad (9.16)$$

where v is the water flow velocity, M is Manning's number describing the surface roughness, R is water flow depth, and I is slope. M varies with different water depth, especially for limited depths on vegetated surfaces (Bengtsson 1997).

Surface roughness can be estimated by the use of remote sensing (land cover classification) and the water depth is estimated by combining precipitation intensity, evapotranspiration, infiltration capacity etc. Slope values will normally be estimated from digital elevation models.

Apart from slope, other topographic parameters like aspect and drainage area are also important in hydrological modelling. The aspect value can, in combination with slope, be used for estimation of irradiance over the surface, and drainage area are commonly used as an independent variable in estimation of soil wetness (Beven and Kirkby 1979). Also the distribution of water over a surface is of course critical in distributed modelling. The following section, based on Pilesjö *et al.* (1998), presents one approach to estimate these parameters.

Catchment topography is critical for models of distributed hydrological processes. Slope controls flow pathways for surface flow, and influences the subsurface flow pattern substantially. The key parameter in catchment topography is flow distribution, which tell us how overland flow is distributed over the catchment area.

Stating flow distribution over a land surface is a crucial measurement in hydrological modelling, the use of Digital Elevation Models (DEM) has made it possible to estimate flow on each location over a surface. Based on the flow distribution estimation on each location represented by a DEM, the drainage pattern over an area, as well as various hydrological parameters, such as catchment area and up-stream flow accumulation, can be modelled.

One common approach for measuring flow distribution is the hydrological flow modelling methods that are widely applied to geomorphological and hydrological problems (Moore *et al.* 1994). This method is based on the following basic principles:

1)   A drainage channel starts from the close neighbourhoods of peaks or saddle points.
2)   At each point of a channel, hydrological flow follows one or more directions of downhill slopes.
3)   Drainage channels do not cross each other.
4)   Hydrological flow continues until it reaches a depression or an outlet of the system.

One critical and most controversial assumption of the hydrological flow modelling method is the determination of flow direction (or drainage path). In the early development, it was assumed that flow follows only the steepest downhill slope. Using a raster DEM, implementation of this method resulted in that hydrological flow at a point only follows one of the eight possible directions corresponding to the eight neighbouring grid cells (Mark 1984; O'Callaghan and Mark 1984; Band 1986; ESRI 1991). Here we call this approach the 'eight-move' algorithm. However, for the quantitative measurement of the flow distribution, this over-simplified assumption must be considered as illogical and would obviously create significant artefacts in the results, as stated by Freeman (1991), Holmgren (1994), Wolock and McCabe (1995), and Pilesjö and Zhou (1996). More complex terrain is supposed to yield more complicated drainage patterns.

Attempts to solve the problem have led to several proposed 'multiple flow direction' algorithms (Freeman 1991; Quinn *et al.* 1991; Holmgren 1994; Pilesjö 1992; Pilesjö and Zhou 1996). These algorithms estimate the flow distribution values proportionally to the slope gradient, or risen slope gradient, in each direction. Holmgren (1994) summarizes some of the algorithms as:

$$f_i = \frac{(\tan \beta_i)^x}{\sum_{j=1}^{8} (\tan \beta_j)^x} \text{ , for all } \beta > 0 \text{ (9.17)}$$

where i,j = flow directions (1...8), f_i = flow proportion (0...1) in direction i, $\tan \beta_i$ = slope gradient between the centre cell and the cell in direction i, and x = variable exponent.

By changing the exponent (x) in Equation (9.17), two extreme approaches in estimating flow distribution can be observed. While $x = 1$, flow will be distributed to downhill neighbouring cells proportionally to the slope gradients, as suggested by Quinn *et al.* (1991). The other extreme is when $x \rightarrow \infty$, which will approach towards the 'eight-move' drainage distribution mentioned above. Holmgren (1994) suggested an x value between 4 and 6. This gives a result between a very homogeneous flow distribution when $x = 1$, and a distinctive flow which occurs when x becomes greater than 10.

This approach focuses on the estimation of flow distribution over a raster DEM. A raster DEM is by far the most commonly used elevation models for modelling purposes, and is easily integrated with other types of spatial data. Given the limitation and problems of the 'eight-move' algorithm, a 'multiple flow direction' approach based on analysis of individual surface facets was taken in this study to estimate flow distribution over the raster DEM. The results from this form-based algorithm were then evaluated and compared with those from the 'eight-move' algorithm.

A number of assumptions when modelling with raster DEM are required in order to isolate the mechanism of overland flow distribution from soil, vegetation and atmospheric impact. The assumptions were:

- The flow from a grid cell to its neighbouring cells is dependent upon the topographic form of the surface.
- Water is evenly distributed over the grid cells (i.e. homogeneous precipitation).
- Every cell in the raster DEM, except sinks, contributes a unit flow and all accumulated flow distributes to its neighbourhood.
- The infiltration capacity over the surface is set to zero.
- The surface is bare (e.g. no vegetation).
- The evaporation is set to zero.
- The topographic form of a cell can be estimated by the use of its eight neighbours (a 3x3 window).

The procedure of a proposed method for the estimation of flow distribution, and subsequently the flow accumulation, from a raster DEM is outlined below:

- Examine each grid cell and the surface facet formed by the 3x3 cell window to classify the facet into 'complicated', 'flat' and 'undisturbed' terrain.
- These three categories are treated separately in determining their flow distribution using the algorithms reported below.
- Results from different categories of the facets are then merged to create flow distribution, and flow accumulation, over the entire study area.
- Classify each facet into 'complicated', 'flat' or 'undisturbed' category.
- The classification of the facet into one of the three categories is based on a normal 3x3 cell filtering process, as outlined below.
- If all the cells in the 3x3 window have the same elevation values the centre cell is marked as a flat surface.
- The eight neighbour cells around the centre cell are examined (clockwise) in order to determine if there are multiple valleys. A valley in this context is defined as one, or more, lower (in elevation) neighbour cells surrounded by cells with higher elevation. The centre cells in these facets are marked as 'complicated' terrain. An example of a facet classified as complicated is presented in Figure 9.8.
- Cells classified as neither 'flat' nor 'complicated' are marked as 'undisturbed'.

.90	.80	.110
.110	.100	.80
.90	.70	.60

Figure 9.8: An example of a surface facet to be classified as 'complicated' terrain. The numbers in the cells denote the elevation values of the centre of the cells. The upper and the upper left cell represent one valley, and the four cells below and right of the centre cell represent another valley (from Pilesjö *et al.* (1998)).

9.4.1 Estimation of topographic form for 'undisturbed' surface facets

There is no absolute way to determine convexity and concavity of the centre cell in a 3 by 3 cell facet. The possible complexity of the surface often implies approximations. One way to approximate is to use a trend surface based on the elevation values of all cells in the facet. Below a method based on a least-squares approximated second-order trend surface (TS) was proposed:

$$TS(x_i, y_i) = a_0 + a_1 x_i + a_2 y_i + a_3 x_i^2 + a_4 y_i^2 + a_5 x_i y_i \quad (9.18)$$

where

$i = 1, \dots , 9$ = The index numbers of the centre cell and its eight neighbours.

$a_0, \dots , a_5$ = The constants for the second-order trend surface.

x_i, y_i = The cell co-ordinates (central cell) in a local system.

Different forms of surfaces should not be modelled in the same way in terms of flow distribution. Complicated terrain, characterized by a number of valleys and ridges, has to be investigated mainly in order to obtain different possible water channels from the centre cell to its neighbours. The flow distribution over flat areas is obviously dependent upon the surrounding topography.

The presence of 'complicated' terrain and flat areas is relatively rare. In most surface facets on a DEM the topography can be classified as 'undisturbed'. This means that the surface is characterized by one single topographic form: concavity or convexity. Convex forms normally occur in the higher parts of the terrain, while the valleys are characterized by concave topographic forms. Since the flow distribution over a convex surface is divergent, and the flow converges over a concave surface, it seems more appropriate to include the topographic form when estimating flow distribution.

The proposed method has demonstrated its ability to produce a better simulation for flow accumulation compared with the 'eight-move' algorithm which is commonly used in today's GIS. However, the results have to be statistically tested, by applying the new algorithm to mathematically generated surfaces.

Another challenge in future research will be focused on appropriate algorithms for generating drainage network, appropriate treatment of sinks, and introduction of more environmental variables into the hydrological modelling processes.

9.5   ESTIMATION OF SUBSURFACE FLOW

Subsurface flow can be described by the use of Darcy's law:

$$q = K * I \qquad (9.19)$$

where q is water flow velocity, K is hydraulic conductivity, and I is the pressure gradient. Since water only can flow in pores filled with water, the flow velocity in these pores, v_p, is

$$v_p = q / \theta \qquad (9.20)$$

where θ is soil moisture. At a point at a specific level, z, the water pressure is ψ.

The total pressure (H), or height of water table, is then

$$H = z + \psi \qquad (9.21)$$

Subsequently, the pressure gradient $I = -dH/dz$ incorporated in Darcy's equation gives:

$$q = -K(1 + d\psi / dz) \qquad (9.22)$$

Subsurface flow can be divided into unsaturated flow in the unsaturated zone and groundwater flow in the saturated zone (see Figure 9.2). When modelling hydrology, the unsaturated flow is often generalized to the vertical dimension, while the saturated flow is modelled in three dimensions. The following two sections are based on Bengtsson (1997).

9.5.1 Estimation of subsurface unsaturated flow

Estimation of unsaturated subsurface flow is normally based on Richard's Equation. The difference between incoming and outgoing flow in a soil with a vertical extension dz corresponds to a difference in soil moisture, θ, in the soil volume during the time dt as:

$$\frac{d\theta}{dt} + \frac{dq}{dz} + e = 0 \qquad (9.23)$$

where e = water loss to vegetation in the root zone and where z has the same positive direction as the flow q. The flow can be calculated by the use of Darcy's law (see above). If we exclude e from Equation 9.23, and substitute q according to Darcy's law we get:

$$\frac{d\theta}{dt} = \frac{d}{dz}\left(K + K\frac{d\psi}{dz} \right) \quad (9.24)$$

Since hydraulic conductivity as well as water pressure depends on soil moisture it is preferable to rewrite equation 9.24 in a way that makes θ the only unknown variable. This can be done with a new variable, D, defined as $D = K*d\psi/d\theta$. We then get:

$$\frac{d\theta}{dt} = \frac{d}{dz}\left(D\frac{d\theta}{dz} + K \right) \quad (9.25)$$

If we know how D and K vary with θ we can then solve the equation numerically.

However, if the soil profile is homogeneous the soil moisture changes rapidly and discontinuously with depth. It is then easier to handle an equation including water pressure instead of soil moisture, since water pressure changes continuously in the profile. This is possible if we substitute with 'specific soil water capacity', defined as $C(\psi) = d\theta/d\psi$. We then get Richardson's Equation as:

$$C\frac{d\psi}{dt} = \frac{d}{dz}\left(K\frac{d\psi}{dz} + K \right) \qquad (9.26)$$

To solve Equation 9.26 we normally have to work with numerical methods. Besides C and K we also need knowledge about initial values along the soil profile and the water pressure above and below the profile. At the soil surface the pressure can be calculated as a function of the infiltration rate, f, or evaporation rate, $-f$, according to:

$$\frac{d\psi}{dz} = \frac{f}{K} - 1 \quad (9.27)$$

If the surface is saturated the water pressure is equal to the water depth. Below the soil profile, at the water table, water pressure is zero.

9.5.2 Estimation of subsurface saturated flow

The groundwater flow is three-dimensional, and the permeable material is often heterogeneous. The hydraulic properties of the ground are characterized by anisotropy, and the hydraulic conductivity is different in different directions. The flow can be estimated by using Darcy's law:

$$v = K * I \quad (9.28)$$

where v is flow velocity and K is saturated hydraulic conductivity and the pressure gradient I is defined as:

$$I = -\frac{dH}{dx} \quad (9.29)$$

where H is total hydraulic pressure height and x is co-ordinates in the flow direction. As mentioned above, Darcy's law is only valid for relatively slow flow, and is not applicable in cracks and in material coarser than gravel.

Since the water cannot flow through the soil mineral particles, but only in the pores between them, the velocity of the water particles, v_p, is faster than flow velocity according to:

$$v_p = v/n \quad (9.30)$$

where n is the soil porosity.

In distributed hydrological modelling, the water table is normally estimated by the use of a DEM. If the distance from the surface to the groundwater is known for every cell, an initial water table can be estimated. When this is done the saturated flow and the fluctuations in the groundwater table can be modelled according to Darcy's law. By comparison of total water pressure in each cell with its neighbours, and with knowledge about hydraulic conductivity in different directions, hydrological conditions can be modelled in time and space.

9.6 SUMMARY

We have given an overview of problems and solutions related to the development of spatially distributed hydrological models. There is a severe gap concerning spatial and temporal scales between the processes we need to mimic and the availability of data. Remote sensing is playing an increasing role in filling this gap, but a number of critical research themes exist where the knowledge is not yet adequate. One such important field is the parameterization of the vegetation's role in the evapotranspiration. Interesting approaches based on the use of thermal remote sensing are under way, but there is need for much more research and improved sensing systems. Another such important field is on modelling the transfer of water between the saturated and the unsaturated zone. One thing we are confident about is that modelling in a GIS environment, with a strong link to remote sensing, is a promising way to go.

9.7 REFERENCES

Abbot, M.B. and Refsgaard, J.C. (eds), 1996, *Distributed Hydrological Modelling.* Dordrecht, Kluwer Academic Publishers.

Andersson, M.G. and Burt, T.P., 1985, *Hydrological Forecasting.* Chichester, John Wiley and Sons.

Andersson, U. and Nilsson, D., 1998, *Distributed Hydrological Modelling in a GIS Perspective – An evaluation of the MIKE SHE Model.* Dept. of Physical Geography, Lund University.

Asrar, G. (ed.), 1989, *Theory and application of optical remote sensing,* New York, Wiley.

Band, L.E., 1986, Topographic partition of watersheds with digital elevation models. *Water Resources Research,* **22**, 15–24.

Begue, A., 1993, Leaf area index, intercepted photosynthetically active radiation and spectral vegetation indices: a sensitivity analysis for regular-clumped canopies. *Remote Sensing of Environment,* **46**, 45–59.

Bengtsson, L., 1997, *Hydrologi – teori och processes.* In Swedish. Svenska Hydrologiska Rådet. Institutionen för Tekn Vattenresurslära. Lund University.

Bergström, S., 1992, *The HBV-model, its structure and applications.* SMHI report (Hydrology) 4.

Beven, K.J. and Kirkby, M.J., 1979, A physically-based, variable contributing area model of basin hydrology. *Hydrological Science Bulletin,* **24**, 1–10.

Beven, K.J. and Moore, I.D., (eds), 1993, *Terrain analysis and distributed modelling in hydrology.* Chichester , John Wiley and Sons.

Brady, N.C. and Weil, R.R., 1996, *The nature and properties of soils.* New Jersey , Prentice-Hall Inc.

Budyko, M.I., 1969, The effect of solar radiation variations on the climate of the Earth. *Tellus,* **21**, 611–1044.

Burrough, P.A. and McDonnell, R.A., 1998, *Principles of Geographical Information Systems.* Oxford, Oxford University Press.

Calder, I.R., 1986, A stochastic model of rainfall interception. *Journal of Hydrology,* **89**, 65–71.

Chorley, R.J., 1977, *Introduction to Physical Hydrology.* London, Methuen and Co. Ltd..

ESRI, 1991, *Cell-based Modelling with GRID,* Environmental System Research Institute, Redlands, CA.

Freeman, T.G., 1991, Calculating catchment area with divergent flow based on a regular grid. *Computers & Geosciences*, **17**, 413–422.

Gash, J.H.C., 1979, An analytical model of rainfall interception by forests. *Quaterly Journal of Roy. Met. Soc.*, **105**, 43–55.

Grip, H. and Rhode, A., 1994, *Vattnets Väg från Regn till Bäck*. In Swedish. Forskningsrådens Förlagstjänst, Uppsala.

Guyot, G., 1998, *Physics of the environment and climate*. Chichester, John Wiley and Sons.

Holmgren, P., 1994, Multiple flow direction algorithms for runoff modelling in grid based elevation models: An empirical evaluation. *Hydrological processes*, **8**, 327–334.

Huete, A.R., 1989, *Soil influences in remotely sensed vegetation-canopy spectra*. Chapter 4 in Asrar 1989.

Jackson, R.D., 1985, Evaluating evapotranspiration at local and regional scales. *Proceedings IEEE*, **73**, 1086–1095.

Jones, J.A.A., 1997, *Global hydrology – Processes, resources and environmental management*. Edinburgh, Longman, 390 pp.

Kirby, M.J., Naden, P.S., Burt, T.P. and Butcher, D.P., 1993, *Computer simulation in Physical Geography*. Chichester, John Wiley and Sons.

Kustas, W.P., Jackson, R.D. and Asrar G., 1989, *Estimating surface energy balance components from remotely sensed data*. In: Asrar, 1989.

Kuzmin, P.O., 1961, Hydrophysical investigations of land waters. International Science Hydrology, *International Union of Geodesy and Geophysics*, **3**, 468–478.

Lind, M. and Fensholt, R., 1999, The spatio-temporal relationship between rainfall and vegetation development in Burkina Faso. *Danish Journal of Geography*, 1999, **2**, 43–56.

Lindroth, A., 1984, *Seasonal variation in pine forest evaporation and canopy conductance*. Avhandling 758, Uppsala Universitet.

Lindström, G., Gardelin, M., Johansson, B., Persson, M. and Bergström, P., 1996, *HBV-96-En Areellt Fördelad Modell för Vattenkrafthydrologin*. In Swedish. SMHI RH Nr. 12, SMHIs tryckeri, Norrköping.

Mark, D.M., 1984, Automated detection of drainage networks from digital elevation models. *Cartographica*, **21**, 168–178.

Monteith, J.L., 1981, Evaporation and surface temperature. *Quaterly Journal of the Roy. Met. Soc.*, **107**, 1–27.

Monteith, J.L. and Unsworth, M., 1990, *Principles of environmental physics*. London , Arnold, 285 pp.

Moore, I.D., Grayson, R.B. and Ladson, A.R., 1994, *Digital terrain modelling: a review of hydrological, geomorphological and biological applications* in Beven, K.J. and Moore, I.D. (eds), Terrain Analysis and Distributed Modelling in Hydrology, Chichester, John Wiley and Sons, UK, 7–34.

Myneni, R.B. and Williams, S.E., 1994, On the relationship between fAPAR and NDVI. *Remote Sensing of Environment*, **49**, 200–211.

O'Callaghan, J.F. and Mark, D.M., 1984, The extraction of drainage networks from digital elevation data. *Computer Vision, Graphics and Image Processing*, **28**, 323–344.

Oke, T.R., 1995, *Boundary Climate Layers*. Cambridge, University Press.

Otterman, J. and Fraser, R.S., 1976, Earth-atmosphere system and surface reflectivities in arid regions from Landsat MSS data. *Remote Sensing of Environment*, **5**, 247–266.

Penman, H.L., 1948, Natural evaporation from open water, bare soil and grass. *Proceedings of the Roy. Soc. Of London*, A194, 20–145.

Persson, D.A. and Pilesjö, P., 2000, Digital elevation models in precision farming. Sensitivity tests of different sampling schemes and interpolation algorithms for the surface generation. *Proceedings of the 2nd International Conference on Geospatial Information in Agriculture and Forestry*, Lake Buena Vista, Florida, 9 – 12 January 2000. pp. II-214-221.

Pilesjö, P., 1992, *GIS and Remote Sensing for Soil Erosion Studies in Semi-arid Environments*, PhD thesis, Meddelanden fran Lunds Universitets Geografiska Institutioner, Avhandlingar CXIV.

Pilesjö, P. and Zhou, Q., 1996, A multiple flow direction algorithm and its use for hydrological modelling in *Geoinformatics '96* Proceedings, April 26-28, West Palm Beach, Florida, 366–376.

Pilesjö, P. and Zhou, Q., 1997, Theoretical estimation of flow accumulation from a grid-based digital elevation model, in *Proceedings of GIS AM/FM ASIA '97 and Geoinformatics '97 Conference*, 26-29 May, Taipei, 447–456.

Pilesjö, P., Zhou, Q. and Harrie L., 1998, Estimating Flow Distribution over Digital Elevation Models using a Formed-Based Algorithm. *Geographic Information Sciences*, **4**, 44–51.

Pinter, P.J., 1992, Solar angle independence in the relationship between absorbed PAR and remotely sensed data for alfalfa. *Remote Sensing of Environment*, **46**, 19–25.

Priestly, C.H.B. and Taylor, R.J., 1972, On the assessment of surface heat flux and evaporation using large scale parameters. *Monthly Weather Review*, **100**, 36–55.

Prince, S.D. and Goward, S.N., 1995, Global primary production: a remote sensing approach. *Journal of Biogeography*, **22**, 815–835.

Quinn, P., Beven, K., Chevallier, P. and Planchon, O., 1991, The prediction of hillslope flow paths for distributed hydrological modelling using digital terrain models. *Hydrological Processes*, **5**, 9–79.

Selby, M.J., 1982, *Hillslope Materials and Processes*. New York, Oxford University Press.

Sellers, P.J., 1992, *Biophysical models of land surface processes in Trenberth 1992*, Climate System Modelling, Cambridge University Press, 780 pp.

Sellers, P.J., Los, S.O., Tucker, C.J., Justice, C.O., Dazlich, D.A., Collats, G.J. and Randall, D.A., 1996, A revised land surface parameterisation (SiB2) for atmospheric GCMs. Part II: the generation of global fields of terrestrial biophysical parameters from satellite data. *Journal of Climate*, **9**, 706–737.

Shaw, E., 1993, *Hydrology in Practice*. London, Chapman and Hall.

Thornthwaite, C.W. and Mather, J.R., 1955, The water balance. Centerton, NJ, Laboratory for climatology publications in *climatology*, **8**, 1–86.

Wanner, W., Strahler, A.H., Muller, J.P, Barnsley, M., Lewis, P., Li, X. and Schaaf, C.L.B., 1995, Global mapping of bi-directional reflectance and albedo for the EOS MODISA project: the algorithm and the product. *Proceedings IGARSS '95*, Firenze,

Wolock, D.M. and McCabe, Jr., G.J., 1995, Comparison of single and multiple flow direction algorithms for computing topographic parameters in TOPMODEL. *Water Resources Research*, **31**, 1315–1324.

10

Remote sensing and geographic information systems for natural disaster management

C.J. van Westen

ABSTRACT

Natural disasters are extreme events within the Earth's system that result in death or injury to humans, and damage or loss of valuable goods, such as buildings, communication systems, agricultural land, forests, natural environment etc. Disasters can have a purely natural origin, or they can be induced or aggravated by human activity. The economic losses due to natural disasters have shown an increase with a factor of eight over the past four decades, caused by the increased vulnerability of the global society due to population growth, urbanization, poor urban planning, as well as an increase in the number of weather-related disasters.

The activities on natural disaster reduction in the past decade, which was designated by the UN as the 'International Decade for Natural Disaster Reduction', have not led to a reduction in these increasing losses. In future even more work has to be done in disaster management. Natural disaster management requires a large amount of multi-temporal spatial data. Satellite remote sensing is the ideal tool for disaster management, since it offers information over large areas, and at short time intervals. Although it can be utilized in the various phases of disaster management, such as prevention, preparedness, relief, and reconstruction, in practice remote sensing is mostly used for warning and monitoring. During the last decades, remote sensing has become an operational tool in the disaster preparedness and warning phases for cyclones, droughts and floods. The use of remote sensing data is not possible without a proper tool to handle the large amounts of data and combine it with data coming from other sources, such as maps or measurement stations. Therefore, together with the growth of the remote sensing applications, Geographic Information Systems (GIS) have become important for disaster management. This chapter gives a review of the use of remote sensing and GIS for a number of major disaster types.

10.1 INTRODUCTION

Natural disasters are extreme events within the Earth's system (lithosphere, hydrosphere, biosphere or atmosphere) which differ substantially from the mean, resulting in death or injury to humans, and damage or loss of 'goods', such as buildings, communication systems, agricultural land, forest, natural environment.

The impact of a natural disaster may be rapid, as in the case of earthquakes, or slow as in the case of drought.

It is important to distinguish between the terms *disaster* and *hazard*. A potentially damaging phenomenon (hazard), such as an earthquake by itself is not considered a disaster when it occurs in uninhabited areas. It is called a disaster when it occurs in a populated area, and brings damage, loss or destruction to the socio-economic system (Alexander 1993). Natural disasters occur in many parts of the world, although each type of disaster is restricted to certain regions. Figure 10.1 gives an indication of the geographical distribution of a number of major hazards, such as earthquakes, volcanoes, tropical storms and cyclones. As can be seen from this figure, earthquakes and volcanoes, for example, are concentrated mainly on the Earth's plate boundaries.

Disasters can be classified in several ways. A possible subdivision is between:

- Natural disasters are events which are caused by purely natural phenomena and bring damage to human societies (such as earthquakes, volcanic eruptions, hurricanes);
- Human-made disasters are events caused by human activities (such as atmospheric pollution, industrial chemical accidents, major armed conflicts, nuclear accidents, oil spills); and
- Human-induced disasters are natural disasters that are accelerated/aggravated by human influence.

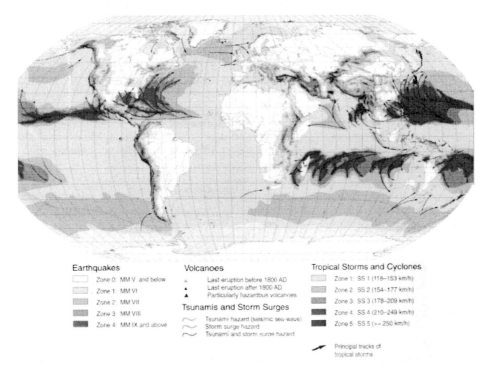

Figure 10.1: World map of natural disasters (Source: Munich Re. 1998).

In Table 10.1, various disasters are classified in a gradual scale between purely natural and purely human-made. A landslide, for example, may be purely natural, as a result of a heavy rainfall or earthquake, but it may also be human induced, as a result of an oversteepened roadcut, or removal of vegetation.

Table 10.1: Classification of disaster in a gradual scale between purely natural and purely human-made.

Natural	Some human influence	Mixed natural / Human influence	Some natural influence	Human
Earthquake	Flood	Landslides	Crop disease	Armed conflict
Tsunami	Dust storm	Subsidence	Insect infestation	Land mines
Volcanic eruption	Drought	Erosion	Forest fire	Major (air-, sea-, land-)
Snow storm /		Desertification	Mangrove decline	traffic accidents
avalanche		Coal fires	Coral reef decline	Nuclear / chemical accidents
Glacial lake		Coastal erosion	Acid rain	Oil spill
outburst		Greenhouse effect	Ozone depletion	Water / soil / air pollution
Lightning		Sea level rise		Groundwater pollution
Windstorm				Electrical power breakdown
Thunderstorm				Pesticides
Hailstorm				
Tornado				
Cyclone/				
Hurricane				
Asteroid impact				
Aurora borealis				

Another subdivision relates to the main controlling factors leading to a disaster. These may be meteorological (too much or too little rainfall, high wind-speed), geomorphological/geological (resulting from anomalies in the Earth's surface or subsurface), ecological (regarding flora and fauna), technological (human made), global environmental (affecting the environment on global scale) and extra terrestrial (See Table 10.2).

The impact of natural disasters to the global environment is becoming more severe over time. The reported number of disasters has dramatically increased, as well as the cost to the global economy and the number of people affected (see Table 10.3 and Figure 10.2).

Earthquakes result in the largest amount of losses. Of the total losses it accounts for 35 per cent, ahead of floods (29 per cent), windstorms (29 per cent) and others (7 per cent). Earthquake is also the main cause in terms of the number of fatalities (48 per cent), followed by windstorms (44 per cent) and floods (8 per cent), (Munich Re. 2001).

The increase in losses and people affected by natural disasters is partly due to the developments in communications, as hardly any disaster passes unnoticed by the mass media. But it is also due to the increased exposure of the world's population to natural disasters. There are a number of factors responsible for this, which can be subdivided into factors leading to a larger risk and factors leading to a higher occurrence of hazardous events. The increased risk is due to the rapid increase of the world population, which has doubled in size from 3 billion in the 1960s to 6 billion in 2000.

Table 10.2: Classification of disasters according to the main controlling factor.

Meteorologi-cal	Geomorphologi-cal/ Geological	Ecological	Technological	Global environmen-tal	Extra terrestrial
Drought Dust storm Flood Lightning Windstorm Thunderstorm Hailstorm Tornado Cyclone/ Hurricane	Earthquake Tsunami Volcanic eruption Landslide Snow avalanche Glacial lake outburst Subsidence Groundwater pollution Coal fires Coastal erosion	Crop disease Insect infestation Forest fire Mangrove decline Coral reef decline	Armed conflict Land mines Major (air-, sea-, land) traffic accidents Nuclear / chemical accidents Oil spill Water / soil / air pollution Electrical power breakdown Pesticides	Acid rain Atmospheric pollution Greenhouse effect Sea level rise El Niño Ozone depletion	Asteroid impact Aurora borealis

Table 10.3: Statistics of great natural disasters for the last five decades
(source: Munich Re 2001).

	Decade 1950 –1959 US $ billion	Decade 1960 – 1969 US $ billion	Decade 1970 – 1979 US $ billion	Decade 1980 – 1989 US $ billion	Decade 1990 – 1999 US $ billion	Last 10 years 1991 – 2000 US $ billion	Factor Last 10: 60s
Number of large disasters	20	27	47	63	89	84	3.1
Economic losses	40.7	73.1	131.5	204.2	629.2	591.0	8.1
Insured losses	0	7.0	12.0	25.5	118.8	104.4	14.9

Depending on the expected growth rates, the world population is estimated to be between 7 and 10 billion by the year 2050 (UNPD 1999).

Another factor related to the population pressure is that areas become settled that were previously avoided due to their susceptibility to natural hazards. Added to this is the important trend of the concentration of people and economic activities in large urban centres, most of which are located in vulnerable coastal areas. Rapidly growing mega-cities mostly occupy marginal land that is more susceptible to disasters.

Another factor related to the increasing impact of natural disasters has to do with the development of highly sensitive technologies and the growing susceptibility of modern industrial societies to breakdowns in their infrastructure. Figure 10.2 shows the distribution of economic and insured losses due to natural disasters during the last 4 decades.

It is also clear that there is a rapid increase in the insured losses, which are mainly related to losses occurring in developed countries. Windstorms clearly dominate the category of insured losses (US $90 billion), followed by earthquakes (US $25 billion). Insured losses to flooding are remarkably less (US $10 billion), due to the fact that they are most severe in developing countries with lower insurance density.

However, it is not only the increased exposure of the population to hazards that can explain the increase in natural disasters. The frequency of destructive events related to atmospheric extremes (such as floods, drought, cyclones, and landslides) is increasing. During the last 10 years a total of 3,750 windstorms and floods were recorded, accounting for two-thirds of all events. The number of catastrophes due to earthquakes and volcanic activity (about 100 per year) has remained constant (Munich Re. 1998). Although the time-span is still not long enough to indicate it with certainty, these data indicate that climate change is negatively related with the occurrence of natural disasters.

There seems to be an inverse relationship between the level of development and loss of human lives in the case of a disaster. About 95 per cent of the disaster-related casualties occur in less developed countries, where more than 4,200 million people live. Economic losses attributable to natural hazards in less developed countries may represent as much as 10 per cent of their gross national product (Munich Re. 1998). In industrialized countries, where warning-systems are more sophisticated, it is more feasible to predict the occurrence of certain natural phenomena, and to carry out mass evacuations. The application of building codes and restrictive zoning also accounts for a lower number of casualties in developed countries.

These statistics illustrate well the importance of hazard mitigation. The International Community has become aware of the necessity to increase the work on disaster management. The decade 1990–2000 was designated the 'International Decade for Natural Disaster Reduction' (IDNDR) by the general assembly of the United Nations. However, now that we are at the end of the IDNDR, we must conclude that the efforts for reducing the effects for disaster reduction during the last decade have not been sufficient.

Great Natural Disasters 1950 - 2000

Far exceeding 100 deaths and/or US$ 100m in claims

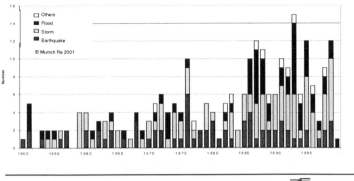

Great Natural Disasters 1950 - 2000

Far exceeding 100 deaths and/or US$ 100m in claims

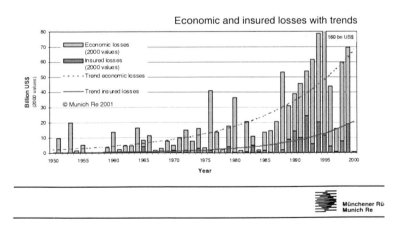

Figure 10.2: Above: number of large natural disasters per year for the period 1950–2000. Below: economic and insured losses due to natural disasters, with trends (Source: Munich Re. 2001).

10.2 DISASTER MANAGEMENT

One way of dealing with natural hazards is to ignore them. In many parts of the world, neither the population nor the authorities choose to take the danger of natural hazards seriously. The complacency may be due to the last major destructive event having happened in the distant past, or people may have moved in the area recently, without having knowledge about potential hazards. Alternatively, the risk due to natural hazards is often taken for granted, given the many dangers and problems confronted by people. Cynical authorities may ignore hazards, because the media exposure and ensuing donor assistance after a disaster has much more impact on voters than the investment of funds for disaster mitigation. To effectively mitigate disasters a complete strategy for disaster management is required, which is also referred to as the disaster management cycle (see Figure 10.3).

Disaster management consists of two phases that take place before a disaster occurs, ***disaster prevention*** and ***disaster preparedness***, and three phases that happen after the occurrence of a disaster, ***disaster relief***, ***rehabilitation*** and ***reconstruction*** (UNDRO 1991). Disaster management is represented here as a cycle, since the occurrence of a disaster will eventually influence the way society is preparing for the next one.

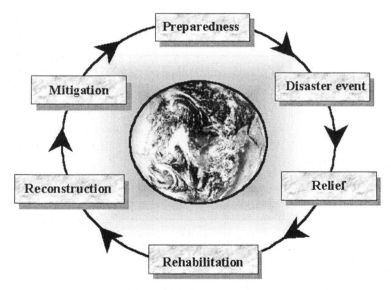

Figure 10.3: The disaster management cycle.

Disaster prevention is the planned reduction of risk to human health and safety. This may involve modifying the causes or consequences of the hazard, the vulnerability of the population or the distribution of the losses. The following activities form part of disaster prevention:

- Disaster preparedness involves all preparatory activities prior to a disaster, so that people can be evacuated, protected or rescued as soon as possible.
- Disaster relief involves the provision of emergency relief and assistance when it is needed and the maintenance of public order and safety.
- Rehabilitation and reconstruction refer to the provision of support during and after a disaster, so that community functions quickly recover.

For more information about disaster management the reader is referred to the following websites:

- The US Federal Emergency Management Agency (FEMA):. The Global Emergency Management System is an online, searchable database containing links to websites in a variety of categories that are related in some way to emergency management. http://www.fema.gov/
- The Office of Foreign Disaster Assistance of the United States Agency for International Development (OFDA/USAID). OFDA also sponsors development of early warning system technology and in-country and international training programs designed to strengthen the ability of foreign governments to rely on their own resources. http://www.info.usaid.gov/ofda/
- The Disaster Preparedness and Emergency Response Association, International (DERA) was founded in 1962 to assist communities world wide in disaster preparedness, response and recovery, and to serve as a professional association linking professionals, volunteers, and organizations active in all phases of emergency preparedness and management. http://www.disasters.org/deralink.html
- Relief Web: a project of the United Nations Office for the Co-ordination of Humanitarian Affairs (OCHA) http://www.reliefweb.int/w/rwb.nsf

10.3 REMOTE SENSING AND GIS: TOOLS IN DISASTER MANAGEMENT

10.3.1 Introduction

Mitigation of natural disasters can be successful only when detailed knowledge is obtained about the expected frequency, character, and magnitude of hazardous events in an area. Many types of information that are needed in natural disaster management have an important spatial component such as maps, aerial photography, satellite imagery, GPS data, rainfall data, etc. Many of these data have different projection and co-ordinate systems, and need to be brought to a common map-basis, in order to superimpose them.

Remote sensing and GIS provide a historical database from which hazard maps may be generated, indicating which areas are potentially dangerous. The zonation of hazard must be the basis for any disaster management project and should supply planners and decision-makers with adequate and understandable information. As many types of disasters, such as floods, drought, cyclones and volcanic eruptions will have certain precursors, satellite remote sensing may detect the early stages of these events as anomalies in a time-series.

When a disaster occurs, the speed of information collection from air and space borne platforms and the possibility of information dissemination with a corresponding swiftness make it possible to monitor the occurrence of the disaster. Simultaneously, GIS may be used to plan evacuation routes, design centres for emergency operations, and integrate satellite data with other relevant data.

In the disaster relief phase, GIS is extremely useful in combination with Global Positioning Systems (GPS) for search and rescue operations. Remote sensing can assist in damage assessment and aftermath monitoring, providing a quantitative base for relief operations.

In the disaster rehabilitation phase, GIS can organize the damage information and the post-disaster census information, as well as sites for reconstruction. Remote sensing updates databases used for the reconstruction of an area.

The volume of data required for disaster management, particularly in the context of integrated development planning, is clearly too much to be handled by manual methods in a timely and effective way. For example, the post-disaster damage reports on buildings in an earthquake stricken city, may be thousands. Each one will need to be evaluated separately in order to decide if the building has suffered irreparable damage. After that all reports should be combined to derive a reconstruction zoning within a relatively short time. GIS may model various hazard and risk scenarios for the future development of an area.

10.3.2 Application levels at different scales

The amount and type of data that has to be stored in a GIS for disaster management depends very much on the level of application or the scale of the management project. Information on natural hazards should be included routinely in development planning and investment project preparation. Development and investment projects should include a cost/benefit analysis of investing in hazard mitigation measures, and weigh them against the losses that are likely to occur if these measures are not taken (OAS/DRDE 1990). Geoinformation can play a role at the following levels:

10.3.2.1 National level

At a national level, GIS and remote sensing can provide useful information, and create disaster awareness with politicians and the public, encouraging the establishment of disaster management organization(s). At such a general level, the objective is to give an inventory of disasters and the areas affected or threatened for an entire country. Mapping scales will be in the order of 1:1,000,000 or smaller. The following types of information should be indicated:

- Hazard-free regions suitable for development;
- Regions with severe hazards where most development should be avoided;

- Hazardous regions where development already has taken place and where measures are needed to reduce the vulnerability;
- Regions where more hazard investigations are required;
- National scale information is also required for those disasters that affect an entire country (drought, major hurricanes, floods etc.).

An example of this application level for the area affected by Hurricane Mitch in 1998 can be found at: http://cindi.usgs.gov/events/mitch/atlas/index.html

10.3.2.2 Regional level

At regional levels the use of GIS for disaster management is intended for planners in the early phases of regional development projects or large engineering projects. It is used to investigate where hazards may constrain rural, urban or infrastructural projects. The areas to be investigated are large, generally several thousands of square kilometres, and the required detail of the input data is still rather low. Typical mapping scales for this level are between 1:100,000 and 1:1,000,000.

Synoptic earth observation is the main source of information at this level, forming the basis for hazard assessment. Apart from the actual hazard information, environmental and population and infrastructural information can be collected at a larger scale than the national level. Thus, GIS can be utilized for analyses at this scale, although the analysis will mostly be qualitative, due to the lack of detailed information.

Some examples of GIS applications at the regional level are:

- Identification of investment projects and preparation of project profiles showing where hazard mitigation measures (flood protection, earthquake resistant structures) should be made.
- Preparation of hazard mitigation projects to reduce risk on currently occupied land.
- Guidance on land use and intensity (OAS/DRDE 1990).

10.3.2.3 Medium level

At this level GIS can be used for the prefeasibility study of development projects, at an inter-municipal or district level. For example for the determination of hazard zones in areas with large engineering structures, roads and urbanization plans. The areas to be investigated will have an area of a few hundreds of square kilometres and a considerably higher detail is required at this scale. Typical mapping scales are in the order of 1:25,000–1:100,000. Slope information at this scale is sufficiently detailed to generate digital elevation models, and derivative products such as slope maps. GIS analysis capabilities for hazard zonation can be utilized extensively. For example, landslide inventories can be combined with other data (geology, slope, land use) using statistical methods to provide hazard susceptibility maps (van Westen 1993).

10.3.2.4 Local level (1:5,000 – 1:15,000)

The level of application is typically that of a municipality. The use of GIS at this level is intended for planners to formulate projects at feasibility levels. But it is also used to generate hazard and risk map for existing settlements and cities, and in the planning of disaster preparedness and disaster relief activities.

Typical mapping scales are 1:5,000– 1:25,000. The detail of information will be high, including for example cadastral information. The hazard data are more quantitative, derived from laboratory testing of materials and in-field measurements. Also the hazard assessment techniques will be more quantitative and based on deterministic/probabilistic models (Terlien *et al.* 1995).

10.3.2.5 Site-investigation scale (> 1:2,000)

At the site-investigation scale GIS is used in the planning and design of engineering structures (buildings bridges, roads etc.), and in detailed engineering measures to mitigate natural hazards (such as retaining walls and checkdams). Typical mapping scales are 1:2,000 or larger. Nearly all of the data are of a quantitative nature. GIS is basically used for the data management, and not for data analysis, since mostly external deterministic models are used for that. Also 3-D GIS can be of great use at this level (Terlien 1996).

Although the selection of the scale of analysis is usually determined by the intended application of the mapping results, the choice of analysis technique remains open. This choice depends on the type of problem, the availability of data, the availability of financial resources, the time available for the investigation, as well as the professional experience of the experts involved in the survey. See also Cova (1999) for an overview of the use of GIS in emergency management.

10.4 EXAMPLES OF THE USE OF GIS AND REMOTE SENSING IN HAZARD ASSESSMENT

10.4.1 Floods

Different types of flooding (e.g. river floods, flash floods, dam-break floods or coastal floods) have different characteristics with respect to the time of occurrence, the magnitude, frequency, duration, flow velocity and the areal extent. Many factors play a role in the occurrence of flooding, such as the intensity and duration of rainfall, snowmelt, deforestation, land use practices, sedimentation in riverbeds, and natural or man has-made obstructions.

Satellite data have been successfully and operationally used in most phases of flood disaster management (CEOS/IGOS 1999). Multi-channel and multi-sensor data sources from meteorological satellites are used for evaluation, interpretation, validation, and assimilation of numerical weather prediction models to assess hydrological and hydro-geological risks (Barrett 1996). Earth observation satellites can be used in many phases of disaster prevention, by mapping geomorphologic elements, historical events and sequential inundation phases, including duration, depth of inundation, and direction of current.

One approach to flood hazard zonation relies on geomorphological analysis of the landforms and the fluvial system, supported wherever possible by information on (past) floods and detailed topographic information. The procedure can be summarized as follows:

- detailed geomorphological terrain mapping, emphasizing fluvial landforms, such as floodplains, terraces, natural levees and backswamps;
- mapping of historical floods by remote sensing image interpretation and field verification to define flooded zone outlines and characteristics;
- overlaying of the geomorphological map and the flood map to obtain indications for the susceptibility to flooding for each geomorphological unit;
- improving the predicting capacities of the method by combination of geomorphological, hydrological, landuse, and other data.

Figure 10.4 shows flood hazard zonation of an area in Bangladesh on reconnaissance (small) scale based on a geomorphological approach to flood hazard mapping using a series of NOAA AVHRR images and a GIS (Asaduzaman *et al.* 1995).

For the prediction of floods, NOAA AVHRR images, combined with radar data, are used to estimate precipitation intensity, amount, and coverage, measure moisture and winds, and to determine ground effects such as the surface soil wetness (Scofield and Achutuni 1996). Quantitative precipitation estimates (QPE) and forecasts (QPF) use satellite data as one source of information to facilitate flood forecasts in order to provide early warnings of flood hazard to communities.

Earth observation satellites are also used extensively in the phases of preparedness/warning and response/monitoring. The use of optical sensors for flood mapping is limited by cloud cover often present during a flood event. Synthetic Aperture Radar (SAR) from ERS and RADARSAT have been proven very useful for mapping flood inundation areas, due to their bad weather capability. In India, ERS-SAR has been used successfully in flood monitoring since 1993, and Radarsat since 1998 (Chakraborti 1999). A standard procedure is used in which speckle is removed with medium filtering techniques, and a piece-wise linear stretching. Colour composites are generated using SAR data during floods and pre-flood SAR images.

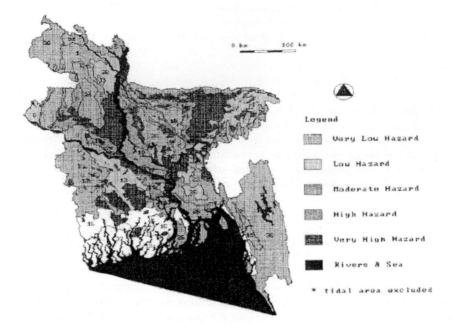

Figure 10.4: Flood hazard zonation map of an area in Bangladesh: results of a reclassification operation using flood frequencies assigned to geomorphological terrain units (Asaduzzaman, 1994)).

For the disaster relief operations, the application of current satellite systems is still limited, due to their poor spatial resolution and problems with cloud cover. Hopefully, higher resolution satellites will improve this (Chapter 3). At a local scale, a large number of hydrological and hydraulic factors can be integrated with high spatial resolution imagery using GIS, especially the generation of detailed topographic information using high precision digital elevation models derived from geodetic surveys, aerial photography, SPOT, LiDAR (Light detection And Ranging) or SAR (Corr 1983). These data are used in two and three dimensional finite element models for the prediction of floods in river channels and floodplains (Gee *et al.* 1990).

10.4.2 Earthquakes

The areas affected by earthquakes are generally large, but they are restricted to well known regions (plate contacts). Typical recurrence periods vary from decades to centuries. Observable associated features include fault rupture, damage due to ground shaking, liquefaction, landslides, fires and floods. The following aspects play an important role: distance from active faults, geological structure, soil types, depth to the water table, topography, and construction types of buildings.

In earthquake hazard mapping two different approaches are to be distinguished, each with a characteristic order of magnitude of map scale (Hays 1980). Small scale (regional) *seismic macro zonation* at scales 1:5,000,000 to 1:50,000 show the likelihood of occurrence and magnitude and the expected recurrence interval of earthquake events in countries or (sub)continents. Large scale (local) *seismic micro zonation* at scales of 1:50-25,000 to 1:10,000 that indicate the magnitude and probability of the various effects of seismic waves at the terrain surface, including ground shaking, surface faulting, tsunamis, landslides and soil liquefaction.

Satellite remote sensing has no major operational role in earthquake disaster management. In the phase of disaster prevention satellite remote sensing can play a role in the mapping of lineaments and faults, the study of the tectonic setting of an area, and neotectonic studies (Drury 1987). Visible and infra-red imagery with spatial resolutions of 5–20 m are generally used.

Satellite Laser Ranging (SLR) and Very Long Base Baseline Interferometry (VLBI) have been used for the monitoring of crustal movement near active faults (see also section 10.4.3). In the measurement of fault displacements GPS have become very important. An increasingly popular remote sensing application is the mapping of earthquake deformation fields using SAR interferometry (InSAR) (Massonet *et al.* 1994, 1996). It allows for a better understanding of fault mechanisms and strain. However, although some spectacular results have been reported, the technique still has a number of problems making routine application difficult.

There are no generally accepted operational methods for predicting earthquakes. Although there is some mention of observable precursors for earthquakes in the literature, such as variations in the electric field or thermal anomalies, they are heavily disputed.

In the phase of disaster relief, satellite remote sensing can at the moment only play a role in the identification of large associated features (such as landslides), which can be mapped by medium detailed imagery such as SPOT and IRS. Structural damage to buildings cannot be observed with the poor resolution of the current systems. The Near Real Time capability for the assessment of damage and the location of possible victims has now become more possible with the availability of the first civilian optical high resolution imagery (IKONOS), though this will only make a difference if adequate temporal resolution, swath-coverage and ready access to the data can be achieved (CEOS/ IGOS 1999).

Unlike remote sensing, the use of GIS in earthquake disaster management is much more prominent. GIS is used in all phases of disaster management. In the prevention phase the large amount of geological, geophysical and geotechnical data are integrated to derive at earthquake response characteristics of soil and the buildings on it.

The use of GIS for earthquake response is even more important. Emergency managers should have adequate knowledge about the extent of the damage shortly after the earthquake. This requires both detailed digital information about the situation before the disaster occurs (for example, about population, buildings infrastructure and utilities), as well as damage assessment models (Emmi and Horton 1995). Also on a longer time frame, damage databases are important for the insurance industry and for the recovery and rehabilitation phase (see Figure 10.5).

Tsunamis are sea waves due to large-scale sudden movement of the ocean floor during earthquakes. They travel at very high speeds in deep ocean waters (up to 900 km/hour), with a long distance (as much as 500 km) between wave crest, and very low heights (around 1 m until the waves approach shallow water. At coastlines, the speed decreases and the wave height increases very rapidly, reaching 25 m or even more. The time interval between waves remains unchanged, usually between 15 minutes and one hour. When tsunamis reach the coastline, the ocean recedes to levels around or lower than the low tide and then rises as a giant destructive wave.

Interferogram   Map showing lahar deposits

Figure 10.5: Left: Interferogram of Mt Pinatubo generated using a tandem pair of ERS images. Right: map showing main deposits related to the 1991 eruption (ash fall, pyroclastic flows) as well as the extension of lahar.

In the Pacific there is an operational Tsunami Warning System (TWS), in the Pacific Warning Center at Honolulu, Hawaii. It consists of 62 tide stations, and 77 seismic stations in 24 countries surrounding the Pacific ocean. The tsunami warnings are effective on a one-hour basis over the Pacific, and 10 minutes on a regional scale. For local tsunamis, systems with a shorter response time are being tested.

10.4.3 Volcanic eruptions

The areas affected by volcanic eruptions are generally small, and restricted to well known regions and may be densely populated. Volcanic eruptions can lead to a diversity of processes, such as explosion (Krakatau, Mount St Helens), pyroclastic flow (Mt Pelee, Pinatubo), lahars (Nevado del Ruiz, Pinatubo), lava flows (Hawaï, Etna), and ashfall (Pinatubo, El Chincon). Volcanic ash clouds can be distributed over large areas, and may impact air-traffic and the weather. Satellite remote sensing has become operational in some of the phases of volcanic disaster management, specifically in the monitoring of ash clouds. The major applications of remote sensing in volcanic hazard assessment are: 1) monitoring volcanic activity and detecting volcanic eruptions, 2) identification of potentially dangerous volcanoes, especially in remote areas and 3) mapping volcanic landforms and deposits (Mouginis-Mark and Francis 1992).

Earth observation satellites can be used in the phase of disaster prevention in the mapping of the distribution and type of volcanic deposits. For the determination of the eruptive history other types of data are required, such as morphological analysis, tephra chronology, and lithological composition. Volcanic eruptions occur within minutes to hours, but are mostly preceded by clear precursors, such as fumarolic activity (gas and smoke emission), seismic tremors, and surface deformation (bulging).

For the (detailed to semi-detailed) mapping of volcanic landforms and deposits, the conventional interpretation of stereo aerial photographs is still the most used technique. The stereo image does not only give a good view of the different lithologies and the geomorphological characteristics of the volcanic terrain, but it can also be used for delineating possible paths of different kinds of lava flows.

One of the most useful aspects of remote sensing is the ability of the visible and infrared radiation to discriminate between fresh rock and vegetated surfaces. This is useful because vegetation quickly develops on all areas except those disturbed by the volcano or other causes (such as urban development). Topographic measurements, and especially the change in topography, are very important for the prediction of volcanic eruptions. Synthetic Aperture Radar (SAR) sensors can provide valuable data that describes the topography. Measurement of ground deformation may eventually be achieved using SAR interferometry.

For the monitoring of volcanic activity a high temporal resolution is an advantage. For the identification of different volcanic deposits a high spatial resolution and, to a lesser extent, also a high spectral resolution are more important.

Hot areas, for example lavas, fumaroles and hot pyroclastic flows, can be mapped and enhanced using Thematic Mapper data. Landsat B and 6 can be used to demonstrate differences in activity which affect larger anomalies such as active block lava flows. For smaller and hotter (>100 C) anomalies the thermal infrared band can be saturated but other infrared bands can be used (Rothery *et al.* 1988; Frances and Rothery 1987; Oppenheimer 1991). Uehara *et al.* (1992) used airborne MSS (1.5 m resolution) to study the thermal distribution of Unzendake volcano in Japan to monitor the lava domes causing pyroclastic flows when collapsed.

Volcanic clouds may be detected by sensors that measure absorption by gases in the cloud such as TOMS (Krueger *et al.* 1994), by infrared sensors such as NOAA AVHRR (Wen and Rose 1994), by comprehensive sensors on meteorological satellites, and by microwave or radar sensors. Remote sensing has become an indispensable part of the global system of detection and tracking of the airborne products of explosive volcanic eruptions via a network of Volcanic Ash Advisory Centers (VAACs) and Meteorological Watch Offices (MWOs). Satellite data provide critical information on current ash cloud coverage, height, movement, and mass as input to aviation SIGnificant METerological (SIGMET) advisories and forecast trajectory dispersion models (CEOS/IGOS 1999).

The assessment of volcanic hazards using GIS is a relatively new approach. Wadge and Isaacs (1988) used GIS techniques to simulate the effects of a wide variety of eruptions of the Soufriere hills volcano, on Montserrat. The energy line concept (Malin and Sheridan 1983) was applied by Kesseler (1995) to model pyroclastic flows, using an energy cone in 3-D. The cone is modelled and compared with a digital elevation model in order to find the potentially affected area. For evaluating the hazard to pyroclastic falls, Kesseler (1995) applied a ballistic model in GIS. Carey and Sparks (1986) presented quantitative models of fallout and dispersal of tephra from volcanic eruption columns. Macedonio *et al.* (1988), and Armeti and Macedonio (1988) made computer simulations for the 79AD Plinian Fall of Mt Vesuvius, and the 1980 tephra transport from Mt St Helens, respectively.

The rapidly changing geomorphology of the watersheds before, during and three consecutive years after the eruption of Mt Pinatubo was investigated by Daag and Van Westen (1996). To quantify the volumes of pyroclastic flow material and the yearly erosion, five digital elevation models were made, and analyzed using a GIS.

Examples of lava flow modelling can be found in Wadge *et al.* (1994). An accurate method is known as the *cellular automata method* (Barca *et al.* 1994). A cellular automata can be considered as a large group of cells with equal dimensions. Each of these cells receives input from its neighbouring cells, and gives output to its neighbours at discrete time intervals. For lava flow modelling, each cell is characterized by specific values (the state) of the following physical parameters: altitude, lava thickness, lava temperature and lava outflow towards neighbouring cells. With this method, the interaction of several lava flows in the same cell can be modelled. Promising results were obtained for the Etna lava flows from 1991–1993. Cellular automata is an example of an inductive-empirical technique, using the taxonomic sheme presented in Chapter 2.

However, although the physical modelling of volcanic processes seems to be a promising and powerful tool, the methods are still in an investigation phase. The results of these models need to be further integrated in a real hazard mitigation project.

10.4.4 Landslides

Individual landslides are generally small in area, but often frequent, in certain mountainous regions. Landslides occur in a large variety of forms, depending on the type of movement (slide, topple, flow, fall, spread), the speed of movement (mm/year - m/sec), the material involved (rock, debris, soil), and the triggering mechanism (earthquake, rainfall, human interaction).

In the phase of disaster prevention, satellite imagery can be used for two purposes: landslide inventory and the mapping of factors related to the occurrence of landslides, such as lithology, geomorphological setting, faults, land use, vegetation and slope. For landslide inventory, mapping the size of the landslide features in relation to the ground resolution of the remote sensing data is very important. A typical landslide of 40000 m^2, for example, corresponds to 20 x 20 pixels on a SPOT Pan image and 10 x 10 pixels on SPOT multispectral images. This would be sufficient to identify a landslide that has a high contrast, with respect to its surroundings (e.g. bare scarps within vegetated terrain), but it is insufficient for a proper analysis of the elements pertaining to the failure to establish characteristics and type of landslide. Imagery with sufficient spatial resolution and stereo capability (SPOT, IRS) can be used to make a general inventory of the past landslides. However, they are mostly not detailed enought to map all landslides. As a consequence, aerial photo-interpretation remains essential.

It is believed that the best airphoto-scale for the interpretation of landslides is between 1:15.000 and 1:25.000 (Rengers *et al.*, 1992); if smaller scales are used, a landslide may only be recognized if size and contrast are sufficiently large. It is expected that in future high-resolution imagery, such as IKONOS, might be used for landslide inventory.

Various methods for landslide hazard using GIS can be differentiated (Van Westen 1993). The most straightforward approach to landslide hazard zonation is a *landslide inventory*, based on aerial photo interpretation, ground survey, and/or a database of historical occurrences of landslides in an area. The final product gives the spatial distribution of mass movements, represented either at scale, as points or as isopleths (Wright *et al.* 1974). In *heuristic methods* the expert opinion of the geomorphologist, making the survey, is used to classify the hazard. The mapping of mass movements and their geomorphological setting is the main input factor for hazard determination (Kienholz 1977; Rupke *et al.* 1988; Hansen 1984).

In statistical landslide hazard analysis, the combinations of factors that have led to landslides in the past are determined statistically and quantitative predictions are made for landslide free areas with similar conditions. In the bivariate statistical method, each factor map (for example slope, geology and land use) is combined with the landslide distribution map, and weight values, based on landslide densities, are calculated for each parameter class (slope class, lithological unit, land use type, etc.). Several statistical methods can be applied to calculate weight values, such as *landslide susceptibility* (Brabb 1984; Van Westen 1993), the *information value method* (Yin and Yan 1988), *weights of evidence modelling Bayesian combination rules, certainty factors, Dempster-Shafer method* and *fuzzy logic* (Chung and Fabbri 1993).

The use of *multivariate statistical* models for landslide hazard zonation has mainly been developed in Italy by Carrara and colleagues (Carrara *et al.* 1990,

1991, 1992). In their applications, all relevant factors are sampled either on a large grid basis, or in morphometric units. Also for each of the sampling units the presence or absence of landslides is determined. The resulting matrix is then analysed using multiple regression or discriminant analysis.

Despite problems related to the collection of sufficient and reliable input data, *deterministic models* are increasingly used in hazard analysis at large scales, especially with the aid of GIS, which can handle the large amount of calculations involved when calculating safety factors. This method is usually applied for translational landslides using the infinite slope model. The methods generally require the use of groundwater simulation models (Okimura and Kawatani 1986). Stochastic methods are sometimes used for selection of input parameters (Mulder 1991; Hammond *et al.* 1992).

10.4.5 Fires

10.4.5.1 Wildfire

The development of a wildland fire depends on three main factors: the fuel (biomass type, condition, moisture etc.), the weather (windspeed, direction, relative humidity, precipitation, temperature) and topography (slope angle, direction, length etc.). Earth observation satellites are used in several phases of fire management such as fuels mapping, risk assessment, detection, monitoring, mapping, burned area recovery, and smoke management.

In the phase of fuel mapping, remote sensing is extensively used to map the vegetation type and vegetation stress. The most frequently used data source for this information is NOAA AVHRR data. Other alternative data sources are ATSR-2; the VEGETATION onboard SPOT 4 as well as the GLI (Global Imager) that will be launched onboard ADEOS-II (CEOS/IGEOS 1999). Existing satellite sensors with wildland fire detection capabilities are under-utilized. They include NOAA-GOES, NOAA-AVHRR, and DMSP-OLS (Robinson 1991).

The TOMS UV aerosol index is used to map the spatial distribution of the UV absorbing aerosols. With the launch of the Tropical Rainfall Measuring Mission (TRMM) program, new remote sensing capabilities now exist to monitor fires, smoke and their impact on the earth-atmosphere system. On the TRMM platform a broadband sensor that measures reflected short-wave radiance in the spectrum between 0.2–4.5 um called the CERES scanner is used. From the UV part of the electro-magnetic spectrum, all the way to the thermal infrared, a combination of sensors can be used to highlight the different features of the smoke and fire events. Each sensor has its own unique capability. From the spatial, spectral and temporal resolution to the number of overpasses during the day, they can provide useful information on the damage to the ecosystems and the impact of fires on the earth-atmosphere system.

For detailed fire assessment, Earth observation satellites such as SPOT and Landsat are currently applied to detect and map burned areas by means of images of a vegetation index (NDVI) based on a specific combination of red and near infrared bands, which specially reflects the amount of green vegetation (Kennedy *et al.* 1994).

Urban encroachment into natural areas, in conjunction with forest and rangeland fire suppression policies, have increased the frequency and intensity of large-area fires in many portions of the world. Similar to flood events, high spatial resolution imagery can be used before, during, and after a fire to measure fuel potential, access, progress, extent, as well as damage and financial loss.

Hamilton *et al.* (1989) discuss the usefulness of GIS for wildfire modelling. They integrated data on topography, weather, and vegetation types, to calculate rate of spread and fireline intensity. Vasconcelos and Guertin (1992) developed the FIREMAP model, which uses GIS spread functions for the calculation of the rate of spread, fireline intensity and direction of maximum spread. The main problems in this model relate to the lack of flexibility of GIS spatial operators and the discrete time nature of the simulations. In order to allow for the modelling of temporal changes in weather and fire conditions, Vasconcels *et al.* (1994) propose the use of distributed discrete event simulation (DEVS) for the spatial dynamic modelling of wildfires.

10.4.5.2 Coal fires

Apart from forest and bush fires, coal fires are one of the largest contributors to CO_2 emissions. In 1992 the CO_2 emission was estimated to be 2–3 per cent of the world's total. Both large underground coal fires occur under natural conditions, as well as in coal mining regions, caused by spontaneous combustion of coal seams.

Remote sensing has proven to be a reliable technology to detect both surface and underground coal fires. A combination of satellite based sensors and airborne sensors are required to unambiguously detect and locate coal fires. By doing such remote sensing based detection on a regular basis, new fires can be detected at an early stage, when they are easier/cheaper to put out. Also, such routine monitoring is very efficient for evaluating the effectiveness of the fire fighting techniques being employed, and which can be remedied/changed as a result.

Thermal infrared data from satellites, especially from the Landsat-5 channel have proven to be very useful. The detection of underground coal fires is limited by the non-uniform solar heating of the terrain. To remove these effects, a DEM can be used for modelling the solar incidence angle. Night-time TM data are more useful for detection, but are not routinely available. On the other hand, due to the low spatial resolution of the TM thermal data (120x120 m) the best night-time TM thermal data can not detect a coal fire less than 50 m even if they have high temperature anomalies. Thus airborne data for detailed detection are still needed.

For the detection and monitoring of coal fires airborne thermal and Landsat TM data have been successfully applied as well as NOAA-AVHRR, ERS1-ATSR and RESURS-01 thermal data (Van Genderen and Haiyan 1997).

10.4.6 Cyclones

Tropical cyclones are intense cyclonic storms which form over warm tropical oceans and threaten lives and property primarily in coastal locations of the tropics, subtropics, and eastern continents in mid-latitudes (World Meteorological Organisation 1995). The term 'tropical cyclones' is often used as a general term for all intensities and locations, including hurricanes, typhoons, tropical storms, tropical depressions, and tropical cyclones. Approximately 80–100 tropical cyclones occur globally in an average year, and their location, season, intensity, and tracks are well documented (WMO 1993).

The following hazards are normally associated with tropical cyclones: strong winds, high ocean waves, flash floods and landslides due to extreme rainfall amounts, and storm surges (which are coastal flooding at the landfall point of the cyclone). In flat areas, the storm surge may reach kilometres far inland and cause considerable loss of life and property damage. The World Meteorological Organisation organizes and co-ordinates tropical cyclone warning centres, in which geostationary satellite imagery is the primary tool for tracking tropical cyclones and estimating intensities. Additional conventional observations, aircraft reconnaissance and numerical model analysis and forecasting, are required for reliable warnings.

Information on the location of the tropical cyclone centres and the way they move can be obtained from weather satellites. The cyclone intensity is estimated using pattern analysis techniques and cloud top temperature information (Dvorak 1984). High-quality animated satellite imagery is used on computer workstations, and objective IR techniques have been developed (CEOS/IGEOS, 1999). Radar images can also provide tracking and intensity information. Doppler radars have been used in locations where tropical cyclones occur.

GIS based emergency systems for cyclone emergency management are used extensively in the south east United States. These provide information including the expected storm surges, which are based on the track and intensity forecast along with a knowledge of the local ocean floor and coastline topography. For evacuation planning, actual evacuation, search and rescue operations and damage assessment, the detailed information stored in the GIS is indispensable.

10.4.7 Environmental hazards

Satellite remote sensing is increasingly being used for mitigation of various environmental hazards. For coastal areas, the accurate location, identification and monitoring of coral reefs, seagrass beds, mangroves, salt marshes, sedimentation, and development activities is greatly facilitated through the use of satellite imagery. Coastal areas can be evaluated for environmental sensitivity and suitability for developing ports, tourist facilities, aquaculture, fisheries etc. Large resolution multispectral imagery can be used for small-scale mapping of wetlands, beaches, submerged vegetation, urbanization, storm damage, and general coastal morphology.

Remote sensing can be used to detect legal and illegal discharges from industrial and municipal facilities into waterways. The surface dimensions of a discharge plume, as well as the source, can be identified and measured if it contains suspended material, such as hydrocarbons, sediments, bubbles, or dye. The effectiveness of containment methods can also be assessed using satellite imagery.

Remote sensing has been used very successfully to detect illegal oilspills by ships on the open sea. In order to be able to prove these illegal acts, it should be possible to trace the ships leaving behind an oil plume, within a short time period.

For this purpose airborne radar systems have been used, but with the disadvantage that only a very limited level of coverage (both spatial and temporal) could be obtained.

Using satellite based SAR, it is possible to detect oil slicks in a wide range of environmental conditions, day and night, at a considerable reduction in cost compared to conventional techniques. At present, satellite SAR data are used within limited areas due to mission constraints on the attainable revisit and spatial coverage. Results from the current exploitation of the data indicate, however, that a greater reliance on satellite data will develop with the new generation of SAR instruments such as the ENVISAT ASAR and Radarsat-2.

Accidental airborne releases of toxic chemicals can be detected and monitored with satellite imagery. For example, if the plume from an oil tank fire is visible to the naked eye, satellite imagery can measure the extent and dissipation of the airborne release, as well as pinpoint the source and identify potential areas of impact downwind.

As one of the purely human-made disasters, landmines may be the most widespread, lethal, and long-lasting form of pollution we have yet encountered. At present, about 10 million anti-personnel mines per year have been produced. The production costs of an anti-personnel mine are between 3 to 30 US dollars. The cost to remove a mine is about 300 to 1000 US dollars. In order to accelerate the mine clearance process new demining methods are urgently required. The military has developed remote sensing techniques to detect minefields. Since the need for humanitarian demining has increased, many of these sensors and techniques are now also available to detect minefields from commercially available platforms and sensors. Attempts have been made to detect minefields by combining the results of several airborne remote sensing sensors which are used on test fields. The sensors used cover the optical, infra-red (thermal) or microwave (radar) region of the electro-magnetic spectrum.
(http://www.itc.nl/ags/research/posters/minefields_general.htm).

10.5 CONCLUSIONS

The decade of the 1990s, designated as the International Decade for Disaster Reduction, has not resulted in a reduction of the losses due to natural disasters. On the contrary, statistics show a rapid increase, both related to an increasing vulnerability of a large part of the Earth's population, as well as to an increase in the number of weather related events. However, there has been a commensurate increase in the technological capabilities and tools that can be used in disaster management. Some of these tools deal with the collection and management of spatial data, such as remote sensing and GIS. Although no satellite was specifically designed to be used in disaster mitigation, most have demonstrated their usefulness in disaster prevention, preparedness and relief.

For several types of disasters, the use of Earth observation techniques has become operational in the warning and monitoring phases for cyclones, drought, and to a lesser extend floods. The operational applications mainly use imagery with low spatial resolution, coming from meteorological satellites or NOAA-type satellite.

For many weather related disasters, obtaining cloud-free images for damage assessment is often impossible. For some types of disasters, such as floods, debris flows or oil spills, SAR is a proven solution. For other types of disasters (e.g. landslides, earthquakes) detailed optical images should be used.

In the phase of disaster relief, satellite remote sensing can only play a role in the identification for large (affected) areas. For example, structural damage to buildings cannot be observed with the poor resolution of the current systems. Near Real Time damage become theoretically possible with the availability of the first civilian optical high-resolution imagery (IKONOS), though this will only make a difference if adequate temporal resolution, swath-coverage and ready access to the data can be achieved. The temporal resolution provided by individual satellites, especially considering cloud cover, will not be sufficient, and high-resolution imagery will not become operational in damage mapping unless multiple satellites are used. This capability is of prime concern to relief agencies requiring near-real time imagery to locate possible victims and structures at risk, and also to map any changes to access that may have occurred. In most cases, the availability of GIS databases, containing information about elements at risk, if combined with less detailed images showing the extent of the area affected, will allow for a rapid assessment of the number of persons and buildings affected.

In view of the limitations inherent to the data collection and analysis techniques and the restrictions imposed by the scale of mapping, especially the phase of hazard assessment within the disaster management cycle will always retain a certain degree of subjectivity. This does not necessarily imply inaccuracy. The objectivity, and certainly the reproducibility, of the assessment can be considerably improved by the interpretation of sequential imagery, a quantitative description of the factors considered, as well as defined analytical procedures and decision rules. The most important aspect however remains the experience of the analyst, both with regard to various factors involved in hazard surveys, as well as in the specific conditions of the study area. Due to the difficulty of formalizing expert rules, the use of expert systems in hazard assessment is still not advanced (Pearson *et al.* 1991).

10.6 REFERENCES

Alexander, D., 1993, *Natural disasters*. University College, London, UCL Press Ltd.

Armeti, P. and Macedonio, G., 1988, A numerical model for simulation of Tephra Transport and Deposition: Applications to May 18, 1980, Mt. St. Helens Eruption. *Journal of Geophysical Research*, **93**, 6463 –6476.

Asaduzzaman, A.T.M., 1994, A geomorphic approach to flood hazard assessment and zonation in Bangladesh, using Remote Sensing and Geographical Information Systems. ITC-MSC thesis. ITC, Enschede, The Netherlands.

Barca, D., Crisci, G.M., Di Gregorio S. and Nicoletta, F.P., 1994, Cellular automata methods for modelling lava flow: a method and examples of the Etnean eruptions. *Transport theory and statistical physics*, **23**, 195–232.

Barrett, E.C., 1996, The storm project: using remote sensing for improved monitoring and prediction of heavy rainfall and related events. *Remote Sensing Reviews*, **14**, 282.

Brabb, E.E., 1984, Innovative approaches to landslide hazard and risk mapping. *Proceedings 4th International Symposium on Landslides*, Toronto, Canada,.vol. 1, 307–324.

Carey, S. and Sparks, R.S.J., 1986, Quantitative models of the fallout and dispersal of tephra from volcanic eruption columns. *Bull. Volcanol.*, **48**, 109–125.

Carrara, A., Cardinali, M. and Guzzetti, F., 1992, Uncertainty in assessing landslide hazard and risk. *ITC Journal*, **1992–2**, 172–183.

Carrara, A., Cardinali, M., Detti, R., Guzzetti, F., Pasqui, V. and Reichenbach, P., 1990, Geographical Information Systems and multivariate models in landslide hazard evaluation. ALPS 90 Alpine Landslide Practical Seminar. *6th International Conference and Field Workshop on Landslides*. August 31 – September 12, 1990, Milano, Italy, pp. 17–28.

Carrara, A., Cardinali, M., Detti, R., Guzzetti, F., Pasqui, V. and Reichenbach, P., 1991, GIS techniques and statistical models in evaluating landslide hazard. *Earth Surface Processes and Landforms*, **16**, 427–445.

CEOS/IGOS, 1999, *CEOS/IGOS disaster management support project*. http://www.ceos.noaa.org/

Chakraborti, A.K., 1999, Satellite remote sensing for near-real-time flood and drought impact assessment – Indian experience. *Workshop on Natural Disasters and their Mitigation – A Remote Sensing and GIS perspective*, 11 – 15 October 1999, Dehradun, India.

Chung, C.F. and Fabbri, A.G., 1993, The representation of geoscience information for data integration. *Nonrenewable Resources*, **2**, 122–139.

Corr, D., 1983, Production of DEM's from ERS-1 SAR data. *Mapping awareness*, **7**, 18–22.

Cova, T.J., 1999, GIS in Emergency management. In: Longley, P.A., Goodchild, M.F. Maguire, D.J., and Rhind, D.V. (ed.). *Geographical Information Systems, Volume 2: Management and Applications*. New York, Wiley & Sons.

Daag, A. and Van Westen, C.J., 1996, Cartographic modelling of erosion in pyroclastic flow deposits of Mount Pinatubo, Philippines. *ITC Journal*, **1996–2**, 110–124.

Drury, S.A., 1987, Image interpretation in geology. London, Allen and Unwin.

Dvorak, V.F., 1984, *Tropical cyclone intensity analysis using satellite data*. NOAA Technical Report NESFIS 1. Washington DC, U.S. Dept. of Commerce.

Emmi, P.C. and Horton, C.A., 1995, A Monte Carlo simulation of error propagation in a GIS-based assessment of seismic risk. *International Journal of Geographic Information Systems*, **9**, 447–461.

Francis, P. and Rothery, D., 1987, Using the Landsat Thematic Mapper to detect and monitor volcanic activity. *Geology*, **15**, 614–617.

Gardner, T.W., Conners Sasowski, K. and Day, R.L., 1990, Automatic extraction of geomorphometric properties from digital elevation data. *Zeitschrift für Geomorphologie* N.F. Supplement Bands **60**, 57–68.

Gee, D.M., Anderson, M.G. and Baird, L., 1990, Large scale floodplain modelling. *Earth surface Processes and Landforms*, **15**, 513–523.

Hamilton, M.P., Salazar, L.A. and Palmer, K.E., 1989, Geographic information systems: providing information for wildland fire planning. *Fire Technology*, **25**, 5–23.

Hammond, C.J., Prellwitz, R.W. and Miller, S.M., 1992, Landslide hazard assessment using Monte Carlo simulation. *Proceedings 6th International Symposium on Landslides*, Christchurch, New Zealand, Vol. 2, 959–964.

Hansen, A., 1984, Landslide Hazard Analysis. In: Brunsden D. and Prior D.B. (ed.), *Slope Instability*. New York, Wiley and Sons.

Hays, W.N., 1980, Procedures for estimating earthquake ground motions. *U.S. Geological Survey Professional Paper*, **1114**, 77

Hellden, U. and Eklundh, L., 1988, National drought impact monitoring: a NOAA NDVI and precipitation data study of Ethiopia. *Lund Studies in Geography Series C. General, Mathematical and Regional Geography*, No. 15, Sweden.

Henricksen, B.L. and Durkin, J.W., 1986, Growing period and drought and early warning in Africa, using satellite data. *International Journal of Remote Sensing*, **7**, 1583–1608.

ISU, 1993, *GEOWARN: Global Emergency Observation and Warning*. International Space University, Design Project Report. Hunstville, Alabama, US.

Keefer, K. *et al.*, 1987, Real-Time Landslide Warning during heavy rainfall. *Science*, **238**, 921–925.

Kennedy, P.J., Belward, A.S. and J. Grégoire, 1994, An improved approach to fire monitoring in West Africa using AVHRR Data. *International Journal of Remote Sensing*, **15**, 2235–2255.

Kesseler, M., 1995, *Modelisation et cartographie de l'area volcanique de Vulcano (Iles Eoliennes, Italie) par un Systeme d'Information Georeferee*. Publications du Departement de Geologie et Paleontologie. Universite de Geneve, 1–72.

Kienholz, H., 1977, *Kombinierte Geomorphologische Gefahrenkarte 1:10.000 von Grindelwald*. Geographica Bernensia G4, Geographisches Institut Universität Bern, Switzerland.

Krueger, A. *et al.*, 1994, Volcanic hazard detection with the Total Ozone Mapping Spectrometer (TOMS), *U.S. Geological Survey Bulletin*, **2047**, 367–372.

Macedonio, G., Pareschi, M.T., and Santacroce, R., 1988, A numerical simulation of the Plinian Fall Phase of the 79AD Eruption of Vesuvius. *Journal of Geophysical Research*, **93**, 14817–14827.

Macedonio, G. and Pareschi, M.T., 1992, Numerical simulation of some lahars from Mount St. Helens. *Journal of Volcanol. Geotherm. Res.*, **54**, 65–80.

Malin, M.C. and Sheridan, M.F., 1982, Computer-assisted mapping of Pyroclastic Surges. *Science*, **217**, 637–639.

Massonnet, D., Feigl, K.L, Rossi, M. and F. Adragna, 1994, Radar interferometric mapping of deformation in the year after the Landers earthquake. *Nature*, **369**, 227–230.

Mouginis–Mark, P.J. and Francis, P.W., 1992, Satellite observations of active volcanoes: prospects for the 1990s. *Episodes*, **15**, 46–55.

Mulder, H.F.H.M., 1991, *Assessment of landslide hazard*. Nederlandse Geografische Studies. PhD thesis University of Utrecht.

Munich Reinsurance Company, 1998, *World Map of Natural Hazards*. Munich Reinsurance Company, Munich, Germany.

Munich Reinsurance Company, 1999, *A year, a century and a millennium of natural catastrophes are all nearing their end – 1999 is completely in line with the catastrophe trend* – Munich Re review. http://www.munichre.com/

Munich Reinsurance Company, 2001, *Annual Review of natural catastrophes 2000*. Munich Reinsurance Company, Munich, Germany.

OAS/DRDE, 1990, *Disaster, Planning and Development: Managing Natural hazards to reduce loss*. Washington DC, US, Department of Regional Development and Environment. Organization of American States,.

Okimura, T. and Kawatani, T., 1986, Mapping of the potential surface–failure sites on granite mountain slopes. In: Gardiner (ed.). *International Geomorphology*. Part 1, New York, John Wiley.

Oppenheimer, C., 1991, Lava flow cooling estimated from Landsat Thematic Mapper infrared data. The Lonquimay eruption (Chile, 1989). *Journal of Geophysical Resources*, **96**, 21865–21878.

Pearson, E., Wadge, G. and Wislocki, A.P., 1991, An integrated expert system/GIS approach to modelling and mapping natural hazards. *Proceedings European conference on GIS (EGIS)*, 763–771.

Rengers, N., Soeters, R. and Westen, C.J. van, 1992, Remote Sensing and GIS applied to mountain hazard mapping. *Episodes*, **15**, 36–45.

Robinson, J.M., 1991, Fire from Space – Global Fire Evaluation using infrared Remote Sensing. *International Journal of Remote Sensing*, **12**, 3–24.

Rothery, D. *et al.*, 1988, Volcano monitoring using short wavelength infrared data from satellites. *Journal of Geophysical Resources*, **93**, 7993–8008.

Rupke, J., Cammeraat, E., Seijmonsbergen, A.C. and Westen, C.J., van, 1988, Engineering geomorphology of the Widentobel catchment, Appenzell and Sankt Gallen, Switzerland. A Geomorphological inventory system applied to geotechnical appraisal of slope stability. *Engineering Geology*, **26**, 33–68.

Scofield, R.A. and Achutuni, R., 1996, The satellite forecasting funnel approach for predicting flash floods. *Remote Sensing Reviews*, **14**, 251–282.

Soeters, R., Rengers, N. and Van Westen, C.J., 1991, Remote sensing and geographical information systems as applied to mountain hazard analysis and environmental monitoring. *Proceedings 8th Thematic Conference on Geologic Remote Sensing (ERIM)*, April 29 – May 2, 1991, Denver, USA, Vol. 2, 1389–1402.

Speed, V., 1994, GIS and satellite imagery take center stage in Mississippi flood relief. *Geo Info Systems*, **4**, 40–43.

Terlien, M.T.J., 1996, *Modelling spatial and temporal variations in rainfall-triggered landslides*. PhD thesis, ITC Publication, Nr. 32, Enschede, The Netherlands.

Terlien, M.T.J., Asch, Th.W.J. van, and Westen, C.J., van, 1995, Deterministic modelling in GIS-based landslide hazard assessment. In: Carrara, A. and Guzzetti, F. (ed.). *Geographical Information Systems in assessing natural hazards*. Kluwer Academic Publishers, 57–77.

Uehara, S., Kumagai, T. and Yazaki, S., 1992, Thermal observations of Unzendake volcanic activities using airborne MSS. *Proceedings: Workshop on Predicting Volcanic Eruptions and Hazard Mitigation Technology*. March 3 – 6, 1992. Tokyo, Japan, Science and Technology Agency.

UNDRO, 1991, *Mitigating Natural Disasters. Phenomena, Effects and Options*. New York, United Nations Disaster Relief Co-ordinator, United Nations.

UNDP, 1999, United Nations Population Division,*World Populations Prospects*. http://www.undp.org/popin/wdtrends/wdtrends.htm

US Foreign Disaster Assistence, 1993, *Report of Disaster History Database*. US Agency for International Development. Washington DC, Office of US Foreign Disaster Assistance.

Van Genderen, J.L. and Guan Haiyan, 1997, *Environmental monitoring of spontanenous combustion in the North China coalfields*. Report EC DG XII-G:I.S.C. Contract Number: C11*-CT93–0008.

Van Westen, C.J., 1993, *Application of Geographic Information Systems to Landslide Hazard Zonation*. ITC-Publication Number 15, ITC, Enschede, The Netherlands, 245

Varnes, D.J., 1984, *Landslide Hazard Zonation: a review of principles and practice*. Commission on Landslides of the IAEG, UNESCO, Natural Hazards No. 3, 61

Vasconcelos, M.J. and Guertin, D.P., 1992, FIREMAP-simulation of fire growth with a geographic information system. *International Journal of Wildland Fire*, **2**, 87–96.

Vasconcelos, M.J., Pereira, J.M.C. and Zeigler, B.P., 1994, Simulation of fire growth in mountain environments. In: Price, M. and Heywood, D.I. (ed.). *Moutain Environments & Geographic Information Systems*. Taylor & Francis, 168–185.

Viedma, O., Meliá, J., Segarra, D. and García-Haro, J., 1997, Modeling rates of ecosystem recovery after fires by using Landsat TM Data. *Remote Sensing of the Environment*, **61**, 383–398.

Wadge, G. and Isaacs, M.C., 1988, Mapping the volcanic hazards from Soufriere Hills Volcano, Montserat, West Indies, using an image processor. *Journal of Geol. Soc.*, **145**, 541–551.

Wadge, G., Young, P.A.V. and McKendrick, I.J., 1994, Mapping lava flow hazards using computer simulation. *Journal of Geophysic Resources*, **99**, 489–504.

Wen, S. and Rose, W., 1994, Retrieval of particle sizes and masses in volcanic clouds using AVHRR bands 4 and 5. *Journal of Geophysic Resources*, **99**, 5421–5431.

World Meteorological Organisation, WMO, 1993, *Global guide to tropical cyclone forecasting*. WMO/TD-No. 560, TCP-31.

World Meteorological Organisation, WMO, 1995, *Global perspectives on tropical cyclones*. WMO/TD No. 693, TCP-38.

Wright, R.H., Campbell, R.H. and Nilsen, T.H., 1974, Preparation and use of isopleth maps of landslide desposits. *Geology*, **2**, 483–485.

Yin, K.L. and Yan, T.Z., 1988, Statistical prediction model for slope instability of metamorphosed rocks. *Proceedings 5th International Symposium on Landslides*, Lausanne, Switzerland, Vol. 2, 1269–1272.

Land use planning and environmental impact assessment using geographic information systems

Egide Nizeyimana, Gary W. Petersen and Joan C. Looijen

11.1 INTRODUCTION

Land use planning consists of making decisions about the use of land and resources (Food and Agricultural Organization 1993). It is conducted primarily to achieve the best use of land, and its implementation is often driven by current and future people's needs in terms of productivity and environmental sustainability. Land use planning may not be as important in sparsely populated countries and communities. In sparsely populated countries or communities, the land use planning is straightforward and is usually aimed at finding the best locations for each of the potential land uses at hand; in other words, it is reduced to a land evaluation (FAO 1976; Rossiter 1996). In highly populated communities, however, land use planning consists of a more elaborate analysis due to conflicts between competing uses (Brinkman 1994). In this case, planning activities are tailored primarily to making the optimal uses of the limited land and associated resources.

Land use planning may be conducted at different levels or scales of decision-making: national, district, local, farm, or field level. The national level is concerned primarily with national priorities as driven by land use policy and legislation. In some countries, land use planning consists of a multi-year program detailing goals at every step of the action plan. Land use planning at a district level often deals with development projects. Local planning may be carried out at the community or watershed level. Farm-level land use planning involves determining best management practices for different fields in the farm. The land use planning process differs from project to project depending on the goals and availability of data. Most projects involve sequential organization of thoughts, evaluation and comparison of alternative land uses in terms of their suitability and impacts to the environment, and finally design and implementation of the plan. While the Food and Agriculture Organization (FAO) land use planning process (Figure 11.1) emphasizes the regional and nationwide planning, that of the United States Department of Agriculture-Natural Resources Conservation Service (USDA-NRCS) is primarily concerned with conservation planning and nutrient management at the farm level (Figure 11.2).

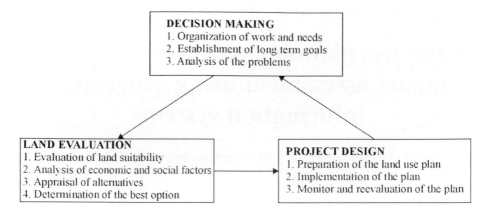

Figure 11.1: Relationships among different tasks of the FAO land use planning process.

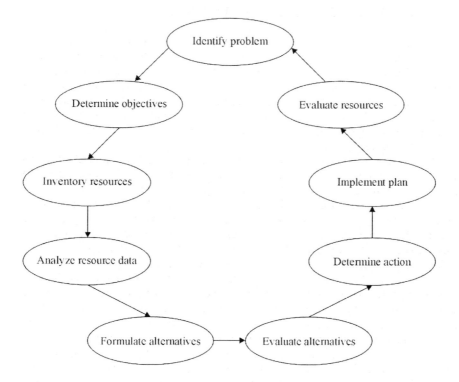

Figure 11.2: The USDA-NRCS farm planning process.

Environmental impact assessment (EIA) is performed in land use planning activities to determine the effect a potential land use would have on the environment (e.g. soil and water pollution, air pollution, ecosystem health and functions, drinking water availability). The type of impact to be evaluated varies depending on the nature of the proposed land use and its potential consequences on the immediate environment. Results of this analysis are intended to help planners to suggest appropriate and sustainable management practices. Only cost-effective land uses that cause the least harm to the environment are retained for further analysis.

Geographic Information System (GIS) technology was originally developed as a tool to aid in the organization, storage, analysis and display of spatial data. The ultimate goal, however, was its application in geographical analysis. Such analyses include land use planning, as well as land use and cover monitoring and management. GIS has been linked to environmental models, decision support systems and expert systems in order to make these tools applicable in spatially-explicit environmental planning and decision-making. The objective of this chapter is to describe ways by which GIS is being or can be used in various aspects of land use planning and associated activities.

11.2 GIS IN LAND USE PLANNING ACTIVITIES

GIS is used for improving the efficiency and effectiveness of a project where geographical information is of prime importance (Burrough 1986). The information within a GIS consists of two elements: spatial data represented by points (e.g. well locations), lines (e.g. streams, road networks), polygons (e.g. soil delineations of soil mapping units), or grid cells, and attribute data or information that describes characteristics of these spatial features. The spatial data are referenced to a geographic spatial coordinate system and are stored either in a vector or raster format (Burrough 1989).

The availability of data sources in a digital form such as digital elevation models (DEM), spatial databases (e.g. soils) and remote sensing imagery, and the increased capability of computers to handle large volumes of data in recent years has increased the applications of GIS in land use planning. In this context, GIS has been used primarily in land evaluation (FAO 1976; Rossiter 1996), a procedure aimed to determine the suitability of land evaluation units for current and alternative uses and the potential impact of each use on the environment.

The development of spatial decision support systems, a type of application software, is one of the reasons GIS is used in land use planning and are defined as interactive computer-based systems that help decision-makers use spatial data and models to solve unstructured problems (Sprague and Carlson 1982). They evolved from decision support system types of decision-making software programs that were developed for business applications including strategic planning, investment appraisal and scheduling of operations (Densham 1991). The goal is to improve the quality of the decision-maker's work, have all the spatial data analysis and modelling capabilities required such that a user, with little or no limited computer and modelling expertise, can quickly access and evaluate the data.

Spatial decision support systems differ from traditional GIS and linear programming models because they provide selection tools via a customized graphical user interface to design and model alternatives, select and evaluate these alternatives, and subsequently display results in maps, tables and/or diagrams. In most straightforward applications, the interface and utility modules are built using the programming facilities offered by various GIS packages. However, spatial decision support systems often combine GIS technology with modelling and programming facilities of computer programming languages such as C, FORTRAN, and Pascal (Petersen *et al.* 1995). For these reasons, GIS have become effective and efficient technologies for scientists, managers and other decision-makers that need to address multi-disciplinary and complex land use planning and management programs. In contrast, GIS-based expert systems provide, not only spatial input data and decision-making information, but also expert knowledge and reasoning rules to manipulate and evaluate that information (van der Vlugt 1989) (see also Chapter 2 – expert system models are classified as deductive-knowledge based models). Unlike a spatial decision support system, an expert system gives potential solutions to the problem.

11.3 SOURCES AND TYPES OF SPATIAL DATA SETS

Data sets commonly used in the land evaluation and EIA can be grouped into two categories: socio-economical and biophysical factors. Socio-economic factors are concerned mainly with the location of the land in relation to essential infrastructures (e.g. markets, major roads, schools, hospitals), social restrictions of the community (e.g. sites of historic value), and economic value of the land use. Biophysical factors, on the other hand, deal with the immediate surface and subsurface environment around the land area of interest. This section deals with sources and types of some of the spatial data commonly used in land evaluation and EIA. Data on soils, vegetation and landform across landscapes are particularly important to GIS-based land use planning activities because they are:

(1)   Often available in GIS format as spatial databases
(2)   Used to provide spatial input data for assessing and modelling the impact of current and alternative land uses on the land, and
(3)   Used to compute land productivity often needed in land evaluations.

11.3.1 Land topography

GIS-based land evaluations can be enhanced by the use of digital topographic data. Contour lines on topographic maps have to be digitized and processed for every area of interest, a tedious and time-consuming process. DEMs are efficient and convenient sources of this information. The US Geological Survey (USGS) produces and distributes DEMs of different resolutions for the entire US The 7.5 minute DEMs are produced from the 7.5-minute quadrangle maps using 30 m grid cells and are referenced horizontally on the Universal Transverse Mercator (UTM) coordinate system (see Chapter 4).

DEMs appear as geo-referenced arrays of regularly spaced elevations of the land surface and are compatible with GIS. Slope gradient and aspect, and curvature (concave, convex and linear) are some of the most important parameters calculated from DEMs (see also Chapter 9) that are relevant to land use planning.

Most GIS packages contain algorithms that automatically compute these topographic characteristics. The slope gradient is one of the key parameters often used in earlier stages of land evaluations to set limits on where a given land use type is possible or not. For example, clean-tilled cropland is generally limited to slopes of less than 18 per cent unless carefully planned management practices are adopted. Slope also affects water infiltration, drainage, storm water runoff, nutrient losses and erosion from croplands. As a result, it is an input data layer to hydrologic, soil productivity, crop growth, and water quality models often used in land evaluations and EIA. Furthermore, a map of slope gradient classes is important to the overall land use planning because it helps locate stable and unstable landscapes for selected potential uses (Marsh 1991). The slope form affects the distribution of water, soils and vegetation and is often used to decide which type of crops should be grown in an area.

11.3.2 Soils

In the US, soil maps have commonly been prepared manually for areas of interest from county soil survey reports. The process is difficult and expensive particularly for large areas, and boundaries between mapping units in these reports are often inaccurate. In addition, thematic consistency is difficult to achieve, because adjacent surveys were typically conducted at different times and with different legends. Today, the process of digitizing soil delineations based on orthophoto quadrangles or topographic maps somewhat corrects these inaccurate boundaries; at least mechanically the boundaries match. This automated processing of soil maps and related attributes allow more efficient handling and display of soil information.

Geographic soil databases are perhaps the most important data source in GIS-based land evaluations. They are developed to organize soil attributes in a coherent manner for efficient storage and handling by users. Early non-digital soil databases consisted of tabular information on soil properties held in a generalized form. An example is the USDA-NRCS map unit interpretation record (MUIR). The use of this type of database, however, is limited due to its poor flexibility when querying data for different, and particularly intense, spatial analyses. Furthermore, soil boundaries have to be redrawn manually every time new data analyses are performed for different interpretative maps (Burrough 1991). Geographic soil databases evolved in order to include both spatial and non-spatial descriptions and were particularly designed to work exclusively within a GIS.

Geographic soil databases have been developed by countries, at a regional or international level or scale. For example, the USDA-NRCS has developed three national soil geographic databases: Soil Survey Geographic (SSURGO), State Soil Geographic (STATSGO) and National Geographic (NATSGO) databases. Each database is composed of soil map units that are linked to attribute tables of chemical and physical properties, and interpretative data (Reybold and TeSelle 1989). SSURGO, the most detailed one, is used primarily in natural resource planning and management at the farm, watershed and county levels. STATSGO was designed for use in statewide and regional environmental assessments (Bliss and Reybold 1989). They are made at the 1:250,000 scale using 1-by 2-degree USGS topographic quadrangles as map bases and are distributed as complete coverage for each state. NATSGO is a broader scale soil database and was designed primarily for regional and national planning. The boundaries between map units are those of the major land resource areas (MLRA) and regions are developed primarily from general state maps. Thus, map units are mainly associations of dominant soil series. The NATSGO soil map was made at a scale of 1:7,500,000 and is distributed as a coverage for the entire US. Similar soil databases have been developed only in a few other countries: Canada, The Netherlands, and France. A major constraint to land evaluation practice in other countries is the lack of reliable digital soil maps.

11.3.3 Land use/cover

Land use and land cover distributions have generally been derived from aerial photographs and land cover maps. Delineations of different land use/cover types on photographs are made from visual interpretations of characteristics (e.g. tone, texture and colour) aided by optical devices. Digital maps of these classes are created by digitizing and processing boundaries between land uses or by scanning photographs covering the area of interest and screen-digitizing their boundaries. These maps have to be geo-referenced and registered before they can be used in GIS analyses. Remote Sensing (RS) imagery is another source of land use/cover information for use in land use planning. Detailed information on satellites presently in orbit is provided in Chapter 3.

While aerial photography and land use maps are effective and appropriate for small area analyses, satellite remote sensing offers tremendous advantages when planning for large areas such as river basins and regions, as it usually allows for relatively easy and reliable differentiation of existing land uses (e.g., wetlands, new subdivisions, degraded landscapes). The knowledge of the extent and location of land cover obtained by classifying remote sensing imagery is crucial to the development of land use plans because it serves as a basis to which other potential land uses can be compared. Land use/cover distributions are input to GIS and GIS-based hydrologic/water quality models commonly used in EIA. They are also important to other aspects of land use planning such as the development and implementation of zoning ordinances, tax assessment, conservation activities, farmland preservation, and ecosystem and wildlife habitat assessments.

Repetitive coverages and the broad-scale perspective of polar-orbiting satellites offer also the possibility of monitoring the land cover changes over time, data which cannot be provided as quickly and efficiently by conventional mapping procedures. Gathering information on land cover at several time intervals is particularly important in the monitoring and re-evaluation stages of a land use planning. Factors such as plant stress, growth and yields that serve as measures of the effectiveness of an existing plan can rapidly be estimated following digital processing and analysis of remote sensing imagery. Remote sensing has been used in many instances to monitor soil degradation and salinity, differentiate geologic formations and locate surface geologic features such as lineaments, etc. (Agbu and Nizeyimana 1991). Although remote sensing cannot fully replace field mapping of land cover and land use, it supplements and provides information not otherwise available to land use planners.

Recent advances in GIS applications have promoted the development of high-resolution spatial technologies that will enhance the land cover assessment and planning in the future. Some of these are the global positioning system (GPS) and digital orthophoto quadrangles (DOQ). GPS allows the user to record accurately and rapidly geographic coordinates of any location in the field with precisions ranging from several tens of metres to a centimetre (Petersen *et al.* 1995). Soil, terrain and land use attributes observed or measured can then be input to a GIS along with their precise locations and extent. Digital orthophoto quadrangles, on the other hand, are digital images of aerial photographs that were corrected to remove relief displacement and distortion caused by the camera angle (Kelmelis *et al.* 1993). Digital orthophoto quadrangles acquired at different times can be used to monitor the magnitude of changes in land cover and land use over time. The relationship between GIS and remote sensing as related to land use planning is illustrated in Figure 11.3.

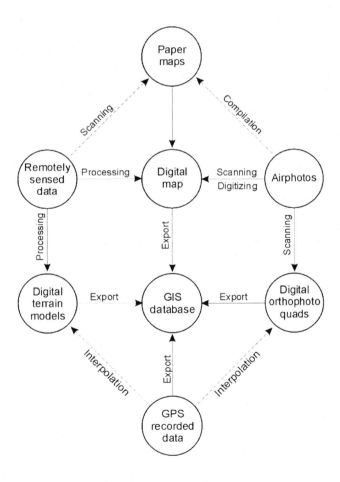

Figure 11.3: Relationships between GIS and remote sensing.

11.4 LAND EVALUATION METHODS

Land evaluation has been defined as the process of assessing or predicting the performance of land for specific purposes (FAO 1976). Potential land uses are analyzed in terms of their ability to provide commodity benefits based on the biophysical, economical, and social factors characterizing the land. Land evaluation is the most important part of the land use planning process (Whitley and Xiang 1993). The land evaluation is performed using conventional or quantitative approaches. In conventional land evaluations, the suitability or limitations for specific uses are generated using ranges of environmental land characteristics including soil properties, landscape-derived data, etc. The resulting classification scheme is often part of national soil surveys and many spatial soil databases. Quantitative approaches, on the other hand, consist of computer-assisted modelling of land qualities. This section deals primarily with biophysical aspects of land evaluation in agriculture and related activities.

GIS is often used in land evaluations for land use planning. In this case, land attributes and qualities are derived from geographic databases and are used to determine land suitability, limitations or ratings for various uses. These databases and associated attribute data may be part of a GIS, a GIS/model linkage, spatial decision support system or expert system. The analysis results may be presented in tabular or graphical form and are intended to provide key information necessary for a land user or planner making meaningful decisions about land management and conservation and/or land use planning. For a theoretical framework of land evaluation approaches, see Rossiter (1996); for a historical survey of land evaluation see van Diepen *et al.* (1991).

11.4.1 Conventional approaches

11.4.1.1 Soil management groupings

Soil management groupings are developed using numerical and descriptive soil information from soil surveys. These interpretations are arranged into classes of relative soil suitability or limitations for a specific use (Soil Survey Division Staff 1993). Several soil interpretative generalizations have been developed for use in agricultural land use planning and management. They are often part of soil survey reports and geographic soil databases. Soil and landscape interpretation systems have been developed in many countries including Great Britain (Morgan 1974), Canada (Canada Land Inventory 1970) and the Netherlands (Vink and van Zuilen 1974). The most well-known examples, however, are perhaps the land capability classification of the USDA-NRCS (Klingebiel and Montgomery 1961) and the fertility capability classification of the FAO (Sanchez *et al.* 1982).

The land capability classification system was developed to assist farmers in planning crop rotations and soil management practices for different fields in the farm (van Diepen *et al.* 1991). In addition to soil properties, the procedure also incorporates information on landform (slope gradient), climate, erosion risk, etc. It consists of eight land capability classes varying from most suitable (I) to least suitable (VIII) soils for agricultural production. Soils occurring in the same class have the same limitations and risks when used for crop growth and therefore should respond similarly to soil conservation and management needs. Each class contains subclasses that indicate the type of limitation (e.g. erosion hazard, wetness). Each subclass may also have several units.

In its original application for conservation farm planning of typical family farms in the US, it has been successful.

The fertility capability classification clusters soils that respond similarly to similar soil management practices (McQuaid *et al.* 1995). However, it is a soil fertility-based ranking system that assigns condition modified according to each soil mapping unit (e.g. low cation exchange capacity, acidity, salinity). Classes are defined according to the number of chemical and physical limitations soils have for proper plant growth. The fertility capability classification is part of the FAO/United Nations Educational, Scientific and Cultural Organization (UNESCO) digital soil database and has been also used in conjunction with country-wide and local soil survey systems.

11.4.1.2 Soil productivity ratings and indices

Soil management groupings described above have been used in some instances as measures of soil productivity (Liu and Craul 1991). However, the productivity of a farm or farm fields for specific crops and plants is commonly evaluated using soil productivity ratings. Soils are compared on a relative scale (0–100) based on predicted yields of a specific crop as a percentage of standard yields taken as 100 per cent (Soil Survey Division Staff 1993). The approach has some advantage since ratings for several crops and different soils can be combined to produce a general rating for a large area. Most county soil survey reports in the US contain soil productivity ratings for major crops and common trees in that county. These data may be entered into a GIS and can be retrieved and used in land evaluations just like any other attribute data.

The level of agricultural productivity for different tracts of a farm can also be determined using ranking/indexing classifications of soil mapping units based on soil properties known to affect crop growth. In this type of classification, soils are ranked in terms of their relative productivity based on a combination of soil properties and qualities relating to productivity potential. As a result, productivity indices are not affected by changes in technology compared to productivity ratings, which are based on plant yields. A soil productivity index model can be multiplicative, additive or a combination of both. In any of these models, a soil parameter is given ranges of values with corresponding ratings that show its suitability for crop production. These are multiplied or added for each mapping unit to create a single index value.

A well-known example of a soil productivity ranking system is the Storie Index (Storie 1937). This index was programmed in FORTRAN as the soil ratings for plant growth (SRPG) model (Sinclair 1996). SRPG is a multiplicative index aimed at classifying soils in terms of their ability to produce fibre, vegetative growth and grains for commodity crops. Soil properties used in the model are derived from soil databases and include organic C, bulk density, clay content, available water capacity, pH, calcium carbonate, CEC, texture, rock fragments, etc. Soil climatic data (moisture and temperature regimes) and landscape features were also incorporated in the final stages of SRPG calculations. The resulting classification is a relative index of land 'quality' for each soil series. Results can be graphically displayed on a farm, county, state or national basis using GIS. This is essentially a multiple-regression approach (i.e. an empirical inductive as described in Chapter 2) to predicting land performance. The resulting apparent precision is attractive, but care must be taken not to over-fit the model (Gauch 1993).

Another example of an often used soil productivity ranking/indexing model is the Productivity Index (Pierce *et al.* 1983). The index rates soils in terms of their suitability for root growth and development. It assumes that crop yields are a function of root growth. The model is a multiplicative function that predicts long-term soil productivity changes due to long-term soil erosion based on soil parameters (pH, available water capacity, bulk density) from the soil interpretation record (SIR) SOILS-5 data set. The model was developed for the north central US region but it can be applied to other areas after proper calibration of its variables.

11.4.2 Quantitative approaches

Quantitative methods used in biophysical land evaluations generally consist of regression and crop growth simulation models. Statistical methods are used to develop empirical models of crop yields (see also Chapter 2). Model coefficients are computed for a specific land location by regressing a sample of yields to corresponding climate (e.g. temperature, rainfall, radiation) and soil characteristics known to affect crop yields. An example of a statistical model relevant to land evaluation is that of Olson and Olson (1986). The authors evaluated various parameters and found that the best fit regression model of yield estimates for areas in the state of New York was a function of rainfall, number of growing degree days, basic cation status and organic C content of soils. Regression models are site-specific and should be calibrated for new land areas. In addition to simple and multiple regression, crop yields may be estimated using more complex statistical approaches such as multivariate and principal component analysis.

During the past few years, scientists have simulated physical and economic land suitability. The modelling approach to land evaluation in land use planning efforts offers some advantages over conventional and statistical methods because the effect and cost associated with temporal changes in land and/or climatic attributes can also be determined. Potential land uses can be compared to crop yields predicted under future climate changes, long-term soil degradation, etc. Similarly, those factors known to affect the socio-economic aspects of land evaluations, such as future population growth and technology changes, can be used to predict the capacity of the land to sustain future food demands.

A number of process-oriented crop simulation models that have potential for use in land evaluation are aimed at simulating crop yields from agricultural lands. The use of models in land evaluations has been very limited particularly due to the large volume of input parameters required and the time involved in compiling them and parameterizing the model, particularly for large areas. The integration of these models with GIS, however, may involve significant processing time.

Sharifi and van Keulen (1994) have developed spatial decision support system for land use planning at the farm level. In this system, planning, crop simulation and linear programming models were integrated into a GIS. A Decision

Support System for Agro-technology Transfer (DSSAT) has been developed by the International Benchmark Sites Network for Agro-technology Transfer which includes many US universities (e.g., University of Hawaii, University of Florida) and the University of Puerto Rico (IBSNAT 1992). The system integrates data from crops, soils, and weather in one standardized input data with several crop models to simulate multi-year yields under different crop management strategies at farm and regional levels. Two GIS-based interfaces, the Agricultural and Environmental Geographic Information System (AEGIS) and AEGIS/WIN were also developed jointly by the University of Florida and University of Puerto Rico and linked to the DSSAT (Engel *et al.* 1993; Luyten and Jones 1997). The AEGIS consists of databases, crop models, ES and GIS (Figure 11.4).

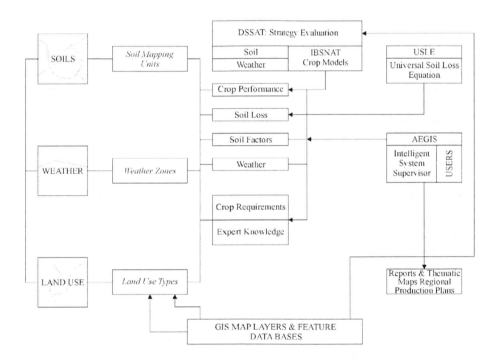

Figure 11.4: AEGIS decision support system (adapted from IBSNAT 1992).

11.4.3 Site suitability analysis

A site suitability analysis typically involves the assessment of the level of affinity a specific land has for a particular land use. Soil information available in soil databases are rarely enough for site evaluations. In addition to soils, the analysis often integrates local information landforms, current land uses, the relative location of the land and associated social and political restrictions. The proposed use may have additional limitations that should be taken into account. For example, an analysis for suitable sites for land application of sewage sludge should, for example, consider the physical, chemical and biological properties of the waste in soils and water.

Sites for various activities have been selected using GIS. Firstly, site-specific analyses often require many and detailed data sources. Secondly, GIS overlay features, logical operations and display functions are tailored to speed up data processing and therefore allow efficiently suitability class assignment and graphical display of results. A good example is the use of GIS for locating appropriate sites for forest land application of sewage waste (Hendrix and Buckley 1992). The authors derived physical site suitability ratings for an area in Vermont based on EPA guidelines (USEPA 1981) and merged them with social and political restrictions of the state and counties to derive a land applicability classification. Similar GIS-based approaches have been used to locate sites for solid waste disposals (Weber *et al.* 1990; Karthikeyan *et al.* 1993).

11.4.4 Standard land evaluation systems

Standard land evaluation systems contain two procedures: biophysical and economic. The biophysical suitability analyses use any of the land capability classification schemes, agricultural productivity ratings or modelling approaches described above. Biophysical land suitability classes may also be derived using land qualities (e.g. erosion hazard, soil water storage) or locally measured/observed land characteristics from sources such as soil databases, remote sensing, geologic maps and DEMs. In this case, numerical values are assigned to suitability ranges of each relevant parameter and land quality. These are combined to yield the overall rating. Suitability classes for ranges of various soil, landscape and hydrologic properties can be found in manuals such as the National Soils Handbook (USDA 1993) or derived by the user based on the knowledge of specific soil behaviour and other land parameters. An economic suitability analysis, on the other hand, includes an evaluation of a land's location in relation to major highways and urban land, and social and economic restrictions of the municipality, county or state.

Many systems have been developed over the years in the agricultural community and in many parts of the world to assist in the land evaluation. An example of such a system is the USDA-NRCS Land Evaluation and Site Assessment (LESA) (Wright *et al.* 1983). LESA was developed to aid planners and public officials at the city, township, county or state level in determining the agricultural quality and economic viability of a farm (Dunford *et al.* 1983; Stamm *et al.* 1987). The land evaluation portion of LESA determines soil productivity levels, farm size and agricultural sales volume; the site assessment portion deals with factors such as location, amount of non-agricultural land, zoning restrictions, etc. (van Horn *et al.* 1989; Daniels 1990). GIS-based LESA systems have also been developed and implemented at county and state levels (Williams 1985; Ferguson *et al.* 1991). LESA and similar systems have been incorporated into many county, state and national programmes in the US, such as the Purchase of Agriculture Conservation Easement programme (PACE) (Christensen *et al.* 1988; Clayville 1994). PACE has been used in farm prioritization programmes such as farmland preservation, taxation and real estate valuation.

A GIS-based land evaluation system, the Comprehensive Resource Inventory and Evaluation System (CRIES), has also been developed (Schultink 1987). The CRIES system uses a grid cell approach to delineate agro-ecological production zones within a GIS and estimates crop production for existing and alternative land uses. It has two inter-linked components: the geographic information system (CRIES-GIS) and the agro-economic information system (CRIES-AIS). The CRIES-GIS component has modules for model parameterization, result display, and other routine GIS-based analyses. The CRIES-AIS has modules for water balance modelling, yield predictions, and linear programming, among others. The system was developed primarily for land use planning applications within the framework of farming systems in developing countries.

Other systems that are commonly used in land evaluations are the Automated Land Evaluation System (ALES) (Rossiter 1990) and the Land Evaluation Computer System (LECS) (Wood and Dent 1983). Both systems are based on the FAO framework of land evaluation. ALES is a PC-based computer program designed to allow land evaluators to build their own expert systems based on the local knowledge of the land. LECS has been implemented in the FAO's Agricultural Planning Toolkit (APT) (Rossiter 1996) and the Integrated Land and Watershed Management Information System (ILWIS) (Elbersen *et al.* 1988) of ITC, The Netherlands. Land evaluation systems may also be part of an SDSS or ES, systems that are designed to aid in the decision-making process.

11.5 LAND USE PLANNING ACTIVITIES AT REGIONAL AND GLOBAL SCALES

Several GIS-based systems have been developed in different countries and regions for land evaluations. In addition to databases, GIS functions and associated programming tools, some systems have modelling capabilities and decision-making tools useful in land resource assessment and evaluations. The FAO Agro-Ecological Zones is an example of such a system (Figure 11.5). Various GIS layers in the agricultural ecological zones are developed from database attributes and combined on a per-grid cell basis to derive model input data for yield simulations, land suitability and productivity evaluations (Koohafkan and Antoine 1997).

Another aspect of regional and global land use planning is the assessment of land quality for use in land and ecosystem assessments and management. Traditional methods for assessing and mapping land soil quality (e.g. soil erosion, soil salinity, and organic matter depletion) are based on observations and laboratory data collected at field and/or watershed scales. Furthermore, measures of composite land qualities such as erosion risk, and the Universal Soil Loss Equation

(USLE), were designed originally to operate at field scales. Today, however, characteristics used in land quality assessments evaluation and subsequently in land use planning can be mapped for large areas using existing digital soil and terrain databases, remotely sensed data and other data sources (Nizeyimana and Petersen 1997). These databases are typically developed by extrapolating field- and watershed-collected data to larger mapping units or by providing links between the spatial data and tables containing interpretation records (Petersen *et al.* 1997).

Results of land quality analyses assist national and international agencies involved in land use planning to understand the regional and global scope of environmental problems such as land degradation (Bliss and Waltman 1994).

A number of countries have developed nationwide geographic soil databases for use in land use planning. In the US, two soil databases described earlier, STATSGO and NATSGO, were designed for use in regional and nationwide environmental resource assessment and planning (Bliss and Reybold 1989), respectively. Soil-based land qualities such as available water storage capacity, organic C storage, acidity, etc., have been mapped using ranges of attribute values from STATSGO (Miller and White 1998) and NATSGO (Kern 1995; Bliss *et al.* 1995) for the entire country. These GIS-derived maps serve to provide a general view of the distribution of limiting or suitable soil attributes to agricultural production in the country. Geographic soil databases have been developed for similar applications in Canada (Kirkwood *et al.* 1996), the Netherlands (DeVries and Denneboom 1993) and Australia (Bui *et al.* 1996). In the latter case, the authors used a regional geographic soil database to determine regional salinization risk resulting from forest clearing in the North Queensland area.

At the continental scale, a soils map of Europe (1:1,000,000 scale) previously compiled and prepared by Tavernier (1985) has been digitized under the Coordinated Information on the European Environment (CORINE) program (Platou *et al.* 1989). The program's objective is the creation of a spatial database on the environment and natural resources of the European communities (Jamagne *et al.* 1996). Its legend is similar to that of the FAO world soil map. Each mapping unit is linked to a soil topological unit that identifies the main soil type whose attribute data are in a relational database. Soil and terrain parameters from this database were used, for example, to generate GIS maps of soil erosion risk for the southern region of the European Community (Bonfils 1989).

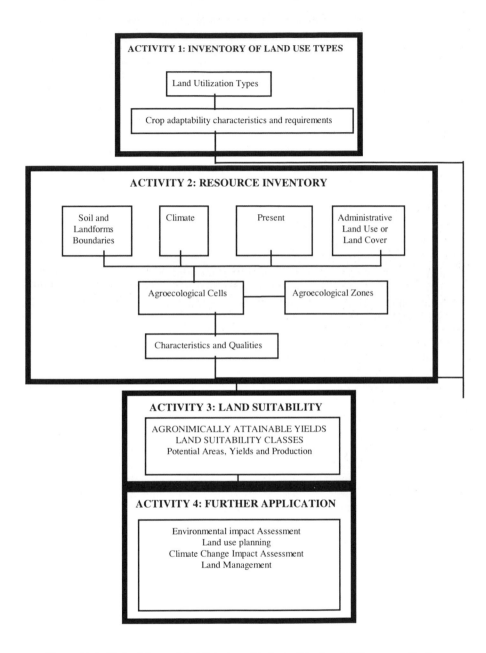

Figure 11.5: Methodology of the FAO Agro-Ecological Zoning (AEZ) core applications (adapted from Koohafkan and Antoine 1997).

At the global scale, a digital soil database has been developed for the entire world (1:5,000,000) by digitizing the FAO/UNESCO soil map of the world (FAO 1995). This GIS coverage consisting of 4930 different map units, has been used to derive land qualities (available water storage, soil productivity potentials, soil degradation potential, etc.) for the US (Kern 1995) and Africa (Eswaran *et al.* 1997). In a similar vein, the development of a more detailed digital database, the World Soils and Terrain (SOTER), has begun under the auspices of the UN/FAO, the International Society of Soil Science (ISSS), and the International Soil Reference and Information Center (ISRIC 1993). SOTER was designed at the 1:1,000,000 scale and accommodates most soil classification systems. Map units in the database are described by one to three terrain components and one to three soils whose characteristics are presented in separate attributes files (Baumgardner 1994).

Owing to the present need for assessing human-induced land degradation, a Global Assessment of Soil Degradation (GLASOD) map that uses SOTER soil and terrain attributes was produced at a scale of 1:10,000,000 through a collaborative project between the United Nations Environment Program (UNEP) and ISRIC (Oldeman *et al.* 1990). GLASOD indicates the geographic distribution of severity of water erosion, wind erosion and physical deterioration such as soil compaction. The information provided by GLASOD is crucial to policy- and decision-makers for establishing priority programs. Other regional and global databases are described in Chapter 4.

11.6 AVAILABILITY AND DISTRIBUTION OF SPATIAL DATA SETS

A number of land information delivery systems have been developed in recent years to make spatial data available to users for application in various aspects of land use planning. These systems are usually designed in a way such that many users can access them at the same time. Some have been linked to the World Wide Web (WWW) for quick and easy access. The data sets may be spatial or in tabular form and are often intended for use in regional or global analyses. An example is the multi-layer soil characteristics data set for the conterminous US (CONUS-SOIL) designed primarily for use in regional climate and hydrology modelling (Miller and White 1998). CONUS-SOIL is based on the STATSGO data and provides soil physical and hydraulic properties including soil textural and rock fragment classes, depth-to-bedrock, bulk density, porosity, rock fragment volume, particle size fractions (sand, silt, and clay), available water storage capacity, and hydrologic soil groups (HSG). Data in CONUS-SOIL are vector or raster (1 km grid resolution) formats and are available for downloading via the Internet or FTP.

During the last few years, scientists have created land resource information systems to develop and update spatial and non-spatial databases and distribute data sets and associated attributes (terrain, land use, etc.) to users. These systems are developed using popular database design software such as INFORMIX and ORACLE, and/or GIS software. GIS-based land resource information systems are multi-purpose systems that integrate geographic databases and GIS tools to analyze, retrieve, record, report and display relationships between data. The type of data in these databases and the complexity of GIS analyses in a land resource information system depends on the intended use of the system. Databases contain data from different sources and scales of application but possess same coordinate system and reference datum for easy integration by users. Table 11.1 shows some of the most recent land resource information systems and their primary functions.

Table 11.1 Most recent land resource information systems and their primary functions.

Land resource information system	Origin	Objectives	Scope	Source
Canadian Soil Information System (CANSIS)	The Centre for Land and Biological Resources Research, Canada	Develop and update spatial soil databases; provide data exchange and distribute digital land resource data	Regional	MacDonald and Valentine (1992)
National Soil Information System (NASIS)	USDA-Natural Resources Conservation Service (NRCS)	Manage and maintain soil data and its dissemination to users	Regional	Soil Survey Division Staff (1991)
ISRIC Soil Information System (ISIS)	International Soil Reference and Information Centre (ISRIC), the Netherlands	Develop soil databases and disseminate soil data	Global	Van der Ven *et al.* (1995)
Global Land Information System (GLIS)	US Geological Survey	On-line descriptive information on the availability status of soils, land cover, terrain data and remote sensing imagery	Global	US Geological Survey (1991)
Land Information Systems (LandIS)	Soil Survey and Land Research Centre, England	Develop and update soil, land cover and related data and its dissemination	Regional	Hallett *et al.* (1996)

11.7   RELIABILITY OF GIS-BASED LAND USE PLANNING RESULTS

GIS analyses for land use planning and EIA have grown in recent years. GIS is particularly attractive in these areas because it allows overlay of spatial data sets and the merging and analysis of attribute data from different sources. The resulting data are obtained using data from digital and paper maps of different scales and/or projections and from data acquired at different resolutions such as in the case of DEMs and remote sensing imagery. The combination of such data layers may produce unrealistic model data and consequently lead to erroneous predictions. The question is how reliable are these results when used for developing land use and management plans. Accuracy of land cover and land use maps, as well as management plans, may be assessed (Skidmore 1999), and the propagation of error and uncertainty in GIS-based analyses is also possible (Heuvelink 1998).

The accuracy of GIS-based land evaluation and EIA is a function of the quality of attribute data and maps, as well as the type of model or assessment scheme used in the analyses. Various algorithms for assessing the quality of GIS results are part of popular GIS and image analysis software (e.g. Arc/Info, ERDAS Imagine), but these tools are frequently not used by GIS practitioners. Similarly, models vary depending on how each represents various processes of the system. Lumped models (see Chapter 2) and indexing/ranking schemes are used mostly in land evaluations because they are easy to parameterize. However, these models ignore spatial variations of parameters throughout the field, watershed or region of study. Furthermore, models originally designed for fields and watersheds are often applied to regional analyses, thus adding some level of uncertainty in modelling predictions. For example, most land use planning programs use conventional methods of land evaluation. Each land parameter is given ranges of values with corresponding ratings showing its suitability to crop production. These indexes are added or multiplied to create a single index that is used to rank land units.

The method is simple but carries a high uncertainty because breaks between two ranges or ranks are subjective.

The effect of map scale and resolution on environmental assessment and modelling output data has been subject to many studies, as illustrated below using DEMs. Raster-based GIS systems require that a grid cell size be defined prior to the analysis. However, as pixel size increases above the resolution of the original data, the spatial variability decreases. This causes a decrease of the predictive power of generated input parameters, particularly for small land areas. An example is the evaluation of DEM cell resolution on hydrologic model input parameters, predicted runoff and peak discharge in watersheds (see also Chapter 9). Studies by Jenson and Domingue (1988) showed that the accuracy and details of watershed boundaries and overland flow paths derived from DEMs depended on the resolution and quality of the DEMs. Chang and Tsai (1991) found that watershed slope gradient decreased gradually as the DEM resolution was varied from 20 to 80 m and differences concentrated in areas of steep slopes. Isaacson and Ripple (1990) reported that slope gradients derived from three arc-second DEM (100 m-DEMs) were lower than those that were determined from 7.5minute DEM for 70 per cent of the total pixel count. Similar results were obtained by Jenson (1991) by comparing ETOPO5, 30 minute-DEMs, and 3 arc-second DEMs. Zhang and Montgomery (1994) and Wolock and Price (1994) showed that increasing the coarseness of the DEM resulted in a decrease in depth to water table and increase in peak flow predicted using TOPMODEL, a topography-based watershed model.

11.8 SUMMARY AND CONCLUSIONS

Environmental scientists provide the information needed to address land use and degradation problems. As the population increases and land becomes a scarce commodity in many parts of the world, carefully planned land uses must be undertaken to accommodate conflicting people's needs and preserve/protect the environment. The decisions about the use of land are made based on analyses of each potential use in terms of its economic and biophysical suitability to the specific tract of land and possible impact to environmental degradation. The definition of the FAO land use planning process has also been extended to include conservation planning and nutrient management planning activities commonly done in the US at farm and watershed levels.

GIS and related technologies (e.g. remote sensing, GPS) have proven to be a valuable tool in land use planning and EIA. The GIS approach is important in these areas because it provides powerful analysis and relational database facilities to modify and/or integrate spatial data from different sources and resolutions. GIS has been coupled to a variety of models (e.g. for crop growth, yield predictions, hydrology and water quality simulations) and is an important component of the spatial decision support system, expert system and land resource information system. In these systems, GIS enhances model flexibility and efficiency. As a result, decision makers and land use planners in local governments, state and regional agencies are using GIS to develop spatial environmental databases, perform land evaluations, and analyze and manage resources.

The demand for GIS and GIS-based analysis systems in land use planning and EIA will increase in the future as more detailed digital environmental data sets become available. Detailed spatial data sets combined with remote sensing, GPS, geostatistics using high capability computers to handle large volumes of data will increase GIS applications in land use planning and management, particularly at scales finer than the farm level.

11.9 REFERENCES

Agbu, A.P. and Nizeyimana, E., 1991, Comparisons between spectral mapping units derived from SPOT image texture and soil map units. *Photogrammetric Engineering and Remote Sensing*, **57**, 397–405.

Baumgardner, M.F., 1994, A world soils and terrain digital database: Linkages. In *Transactions of the 15th World Congress of Soil Science*. Volume 6a: Commission V: Symposia, Acapulco, International Society of Soil Science) pp. 718–727.

Bliss, N.B. and Reybold, W.U., 1989, Small-scale digital soil maps for interpreting natural resources. *Journal of Soil and Water Conservation*, **44**, 30–34.

Bliss, N.B. and Waltman, S.W., 1994, Modelling variations of land qualities at regional and global scales using geographic information systems. In *Transactions of the 15th World Congress of Soil Science*. Volume 6a: Commission V: Symposia. Acapulco, International Society of Soil Science. pp. 644–661.

Bliss, N.B., Waltman, S.W. and Petersen, G.W., 1995, Preparing a soil carbon inventory for the United States using geographic information systems. In *Soils and Global Change*, edited by R. Lal, J. Kimble, E. Levine and B.A. Stewart. Boca Raton, Lewis Publishers, 275–295.

Bonfils, P., 1989, Une evaluation du risqué d'érosion dans les pays du sud de la communauté européene (Le Programme CORINE). *Science du Sol*, **27**, 33–36.

Brinkman, R., 1994, Recent developments in land use planning, with special reference to FAO. In: Fresco, L.O., Stroosnijder, L., Bouma J., and van Keulen

H. (ed.). *The Future of Land: Mobilizing and Integrating knowledge for Land Use Options.* Chichester, John Wiley & Sons, pp. 13–21.

Bui, E.N., Smettem, R.J., Moran, C. and Williams, J., 1996, Use of soil survey information to assess regional salinization risk using a geographic information system. *Journal of Environmental Quality*, **25**, 433–439.

Burrough, P.A., 1986, Principles of Geographical Information Systems for Land Resources Assessment. *Monographs on Soil and Resources*, Survey no. 12. Oxford, Oxford Science Publications.

Burrough, P.A., 1989, Matching databases and quantitative models in land resource assessment. *Soil Use and Management*, **5**, 3–8.

Burrough, P.A., 1991, Soil information systems. In: Maguire, D.J. Goodchild M.F. and Rhind D.W. (ed.). *Geographic Information Systems. Principles and Applications.* Volume 2: Applications. Essex, Longman Scientific & Technical, pp. 99–106.

Canada Land Inventory, 1970, *The Canada land inventory: Objectives, scope and organization.* Report 7, Department of Regional Economic Expansion, Ottawa.

Chang, K. and Tsai, B., 1991, The effect of DEM resolution on slope and aspect mapping. *Cartography and Geographic Information Systems*, **18**, 69–77.

Christensen, G.R., Budd, W.W., Reganold, J.P. and Steiner, F.R., 1988, Farmland protection in Washington State: An analysis. *Journal of Soil and Water Conservation*, **43**, 82–89.

Clayville, B.L., 1994, *Applications of Geographic Information Systems to Pennsylvania's Agricultural Conservation Easement Purchase Program.* MSc thesis. University Park, PA, Department of Agronomy, Pennsylvania State University.

Daniels, T., 1990, Using LESA in a purchase of development rights program. *Journal of Soil and Water Conservation*, **45**, 617–621.

Densham, P.J., 1991, Spatial decision support systems. In Maguire, D.J. Goodchild M.F. and Rhind D.W. (ed.). *Geographic Information Systems, Vol. 1: Principles and Applications.* Essex, Longman Scientific & Technical, pp. 403–412.

DeVries, F. and Denneboom, J., 1993, *The Digital version of the Dutch Soil Map of the Netherlands.* SC-DLO Technical Document 1. Wageningen, Winand Staring Center.

Dunford, R.W., Roe, R.D., Steiner, F.R., Wagner, W.R. and Wright, L.E., 1983, Implementing LESA in Whitman Co., Washington. *Journal of Soil and Water Conservation*, **38**, 87–89.

Elbersen, G.W., Ismangum, W., Sutaatmadja, D.S. and Solihin, A.A., 1988, Small scale soil survey and automated land evaluation. *ITC Journal*, **1988–1**, 51–59.

Engel, B.E., Srinivasan, R. and Rewerts, C., 1993, A spatial decision support system for modelling and managing agricultural nonpoint soure pollution. In: Goodchild, M.F. Parks, B.O. and Steyaert, L.T. (ed.). *Environmental Modelling with GIS.* New York, Oxford University Press, pp 231–237.

Eswaran, H., Almarez, R., Van den Berg, V. and Reich, P., 1997, *An assessment of soil resources of Africa in relation to soil productivity.* http://www.nhq.nrcs.usda.gov/WSR/papers/africa.html. World Soil Resources, Soil Survey Division, USDA-NRCS Washington D.C.

(FAO), 1995, *Digital soil map of the World and derived soil properties*. (Version 3.5). CD-ROM. Food and Agricultural Organization of the United Nations. Rome, FAO.

FAO, 1993, *Guidelines for Land Use Planning*, FAO Development Series 1, Soil Resources, Management and Conservation Service. Rome, FAO.

FAO, 1976, *A framework for land evaluation*. Soils Bulletin no. 55. Food and Agricultural Organization of the United Nations. Rome, FAO.

FAO/CSRC, 1974, *Soil map of the World*. 1:5,000,000. UNESCO, Paris, France. 1/2° Digitization. Durham, Complex Systems Research Center.

Ferguson, C.A., Bowen, R.L. and Kahn, M.A., 1991, A statewide LESA system for Hawaii. *Journal of Soil and Water Conservation*, **46**, 263–267.

Gauch, Jr., H.G., 1993, Prediction, parsimony and noise. *American Scientist*, **81**, 468–478.

Global Soils Data Task, 1995, *Global Change Data sets*. A collaborative project: FAO, ISRIC, IGBP, NRCS. Lincoln, NE.

Hallett, S.H., Jones, R.J.A. and Keay, C.A., 1996, Environmental information systems developments for planning sustainable land use. *International Journal of Geographic Information Systems*, **10**, 47–64.

Hendrix, W.G. and Buckley, J.A., 1992, Use of a geographic information system for selection of sites for land application of sewage waste. *Journal of Soil and Water Conservation*, **92**, 271–275.

Heuvelink, G.B.M., 1998, *Error propagation in environmental modelling with GIS*. London, Taylor and Francis.

IBSNAT Project, 1992, Linking DSSAT to a geographic information system, IBSNAT. *Agrotechnology Transfer*, **15**, 1–6.

International Soil Reference and Information Center (ISRIC), 1993, *Global and National Soils and Terrain Database (SOTER)*. Procedures Manual. World Soil Resources Report no. 74. Rome, FAO.

Isaacson, D.L. and Ripple, W.J., 1990, Comparison of 7.5-minute and 1-degree digital elevation models. *Photogrammetric Engineering and Remote Sensing*, **56**, 1523–1527.

Jamagne, M., King, D., Le Gas, C., Daroussin, J., Burrill, A. and Vossin, P., 1996, Creation and use of a European soil geographic database. In *Transactions of the 15ᵗʰ World Congress of Soil Science*. Volume 6a: Commission V: Symposia. Acapulco, International Society of Soil Science, pp. 728–742.

Jenson, S.K., 1991, Applications of hydrologic information automatically extracted from digital elevation models. *Hydrologic Processes*, **5**, 31–44.

Jenson, S.K. and Domingue, J.O., 1988, Extracting topographic structure from digital elevation data for geographic information system analysis. *Photogrammetric Engineering and Remote Sensing*, **54**, 1593–1600.

Karthikeyan, K.G., Elliott, H.A. and Brandt, R.C., 1993, Using a geographic information system for siting water treatment sludge monofills. In *Proceedings of the International Symposium on Integrated Resource Management and landscape modification for Environmental Protection*, edited by J.K. Mitchell. Saint Joseph, American Society of Agricultural Engineers, 315–325.

Kelmelis, J.A., Kirtland, D.A., Nystrom, D.A. and VanDriel, N., 1993, From local to global scales. *Geo Information Systems*, **4**, 35–43.

Kern, J.S., 1995, Geographic patterns of soil water-holding capacity in the conterminous United States. *Soil Science of America Journal*, **59**, 1126–1133.

Kirkwood, V., Dumanski, J., Bootsma, A., Stewart, R.B. and Muma, R., 1996, *Land Resource Research Centre*. Research Branch, Technical Bulletin 1983-4E. LRRC Contribution no. 86–29. Ottawa, Ontario, Agriculture Canada.

Klingebiel, A.A. and Montgomery, P.H., 1961, *Land Capability Classification, Agriculture Handbook no. 210.* Washington D.C.USDA Soil Conservation Service.

Koohafkan, A.P. and Antoine, J., 1997, *Application of Agro-Ecological zoning and GIS tools for land use planning, food security strategy development and land vulnerability assessments*. Unpublished report, Rome, FAO,.

Liu, R. and Craul, P.J., 1991, A GIS-based soil capability classification for agriculture in Yuisu County, China. In *Annual Convention of the ACSM-ASPRS*. Vol. 4, Technical Papers. Bethesda, American Society for Photogrammetry and Remote Sensing, American Congress on Surveying and Mapping, pp. 99–106.

Luyten, J.C. and Jones, J.W., 1997, AEGIS+: A GIS-based graphical user-interface for defining spatial crop management strategies and visualization of crop simulation results. *Agronomy Abstract*, **89**, 23.

MacDonald, K.B. and Valentine, K.W.G., 1992, CanSIS/NSDB. *A general description*. CLBRR Contribution Number 92–35. Centre for Land and Biological Resources Research Branch, Agriculture Canada, Ottawa.

Marsh, W.M., 1991, *Landscape Planning: Environmental Applications*. New York, John Wiley and Sons.

McQuaid, B.F., Buol, S.W. and Kelley, J.A., 1995, Application of the soil fertility capability classification in soil survey reports. *Soil Survey Horizons*, **36**, 117–121.

Miller, D.A. and White, R.A., 1998, *A conterminous United States multi-layer soil characteristics dataset for regional, climate and hydrology modelling*. Earth Interactions, 2, [Available on line at http://EarthInteractions.org].

Morgan, J.P., 1974, ADAS (Lands) physical agricultural classification. In *Land Capability Classification*, Ministry of Agriculture, Fisheries and Food, Technical Bulletin no. 30, London, HMSO, pp. 80–89.

Nizeyimana, E. and Petersen, G.W., 1997, Remote sensing applications to soil degradation assessments. In: Lal, R. Blum, W.H. alentine C. V. and Stewart B.A. (ed.). *Methods for Assessment of Soil Degradation*. Boca Raton:, CRC Press, pp. 393–405.

Oldeman, L.R., Hakkeling, R.T.A. and Sombroek, W.G., 1990, *World map of the status of Human-Induced soil degradation: An explanatory note*. Global Assessment of Soil Degradation (GLASOD), Wageningen, ISRIC/UNEP.

Olson, K.R. and Olson, G.W., 1986, Use of multiple regression analyses to estimate average corn yields using selected soils and climatic data. *Agricultural Systems*, **20**, 105–120.

Panuaska, J.C., Moore, I.D. and Kramer, L.A., 1991, Terrain analysis: Integration into the Agricultural Nonpoint Source (AGNPS) pollution model. *Journal of Soil and Water Conservation*, **46**, 59–64.

Petersen, G.W., Bell, J.C., McSweeney, K., Nielsen, G.A. and Robert, P.C., 1995, Geographic information systems in agronomy. *Advances in Agronomy*, **55**, 67–111.

Petersen, G.W., Nizeyimana, E. and Evans, B.M., 1997, Application of GIS to land degradation assessments In: Lal, R. Blum, W.H. alentine C. V. and Stewart B.A. (ed.). *Methods for Assessment of Soil Degradation*. Boca Raton:, CRC Press, pp. 377–391.

Pierce, F.J., Larson, W.E., Dowdy, R.H. and Graham, W.E.P., 1983, Productivity of soils: Assessing long-term changes due to erosion. *Journal of Soil and Water Conservation*, **38**, 131–136.

Platou, S.W., Nørr, A.H. and Madsen, H.B., 1989, Digitization of the EC soil map. In: Jones, R.J.A. and Biagi, B. (ed.). *Computerization of Land Use Data*. EUR 11151 EN. Luxembourg, Office for official publications of the European Communities, pp. 132–145.

Reybold, W.U. and TeSelle, G.W., 1989, Soil geographic databases. *Journal of Soil and Water Conservation*, **44**, 28–29.

Rossiter, D.G., 1990, ALES: A framework for land evaluation using a micro computer. *Soil Use and Management*, **6**, 7–20.

Rossiter, D.G., 1996, A theoretical framework for land evaluation. *Geoderma*, **72**, 165–190.

Rossiter, D.G., 1998, *Lecture Notes: Methodology for Soil Resource Inventories* (SOL.27). Soil Science Division, ITC, Enschede, NL. November 1998. On-line reference:http://www.itc.nl/~rossiter/Docs/K5_lectures.pdf/

Rossiter, D.G., 1999, *Lecture Notes: Soil Geographic Data Bases (SOL.SGDB)*. Soil Science Division, ITC Enschede, NL. February 1999. On-line reference: http://www.itc.nl/~rossiter/Docs/SoilGeographicDataBases.pdf/

Sanchez, P.A., Couto, W. and Buol, S.W., 1982, The fertility capability soil classification system: Interpretation, applicability and modification. *Geoderma*, **27**, 283–309.

Schultink, G., 1987, The CRIES resource information system: Computer-aided land resource evaluation for development planning and policy analysis. *Soil Survey and Land Evaluation*, **7**, 47–62.

Sharifi, M.A. and van Keulen, H., 1994, A decision support system for land use planning at farm enterprise level. *Agricultural Systems*, **45**, 239–257.

Sinclair, H.R., 1996, Soil ratings for plant growth. *Agronomy Abstract*, **88**, 279.

Sinclair, H.R., Waltman, W.J., Terpstra, H.P. and Margentan, D.R., 1996, *Soil Ratings for Plant Growth*. USDA-NRCS National Soil Survey Center, Lincoln, Nebraska.

Skidmore, A.K., 1999, Accuracy assessment of spatial information. In: Stein, A., van der Meer, F. and Gorte, B., 1999. *Spatial Statistics for Remote Sensing*. Kluwer Academic Publishers, Dordrecht, Chapter 12, 197–209.

Soil Survey Division Staff, 1991, *National Soil Information System (NASIS)-soil interpretation and information dissemination subsystem*. Draft requirements statement. USDA-NRCS, National Soil Survey Center, Lincoln, Nebraska.

Soil Survey Division Staff, 1993, *Soil Survey Manual*. USDA Agriculture Handbook 18. Washington DC, US Government Printing Office.

Sprague, R.H. and Carlson, E.D., 1982, *Building effective decision support systems*. Englewood, Prentice Hall.

Stamm, T., Gill, R. and Page, K., 1987, Agricultural land evaluation and site assessment in Latah County, Idaho. *Environmental Management*, **11**, 379–388.

Storie, R.E., 1937, *An index for rating agricultural value of soils.* Bull. 556. Berkley, University of California.

Tavernier, R., 1985, *Soil map of the European Communities.* 1:1,000,000-scale. 1 map. Luxembourg, Office of Publications of the European Communities.

US Department of Agriculture (USDA), 1993, *National Soils Handbook.* USDA-SCS. Washington DC, US Government Printing Office.

US Environmental Protection Agency, 1981, *Process design manual for land treatment of municipal wastes.* PB 299655/1. Cincinnati, Environmental Protection Agency (USEPA).

US Geological Survey, 1990, *Digital Elevation Models. Data Users Guide 5.* US Geodata, Technical Instructions. Reston, USGS National Mapping Program.

US Geological Survey, 1991, *Global Land Information System (GLIS), interactive query, data visualization and order system.* Sioux Falls, EROS Data Center.

Van der Ven, T., Tempel, P. and Verhagen, J., 1995, *ISRIC Soil Information System ISIS 4.0. User Manual. Revised edition.* Technical Paper 15 Wageningen, The Netherlands, International Soil Reference and Information Centre.

Van der Vlugt, M., 1989, The use of a GIS based decision support system in physical planning. *Proceedings of GIS/LIS'91. Vol. 1.* Bethesda, Maryland, American Society for Photogrammetry and Remote Sensing/American Congress on Surveying and Mapping, pp. 459–467.

Van Diepen, C.A., Van Keulen, H., Wolf, J. and Berkhout, J.A.A., 1991, Land evaluation: From intuition to quantification. *Advances in Soil Science*, **15**, 150–204.

Van Horn, T.G., Steinhardt, G.C. and Yahner, J.E., 1989, Evaluating the consistency of results for the agricultural land evaluation and site assessment (LESA) system. *Journal of Soil and Water Conservation*, **44**, 615–620.

Vink, A.P. and van Zuilen, E.J., 1974, *Suitability of the soils of the Netherlands for arable land and grassland.* Soil Survey Papers no. 8, Wageningen, Netherlands Soil Survey Institute.

Weber, R.S., Jenkins, J. and Leszkiewics, J.J., 1990, Application of geographic information system technology to landfill site selection. In *Proceedings of the Application of Geographic Information Systems and Knowledge-Based Systems for Land Use Management.* Blacksburg, Virginia Polytechnic Institute and State University, pp. 421–430.

Weir, A.H., Bragg, P.L., Porter, J.R. and Rayner, J.H., 1984, A winter wheat crop simulation model without water and nutrient limitations. *Journal of Agricultural Science*, **102**, 311–382.

Whitley, D.L. and Xiang, W.N., 1993, A GIS-based method for integrating expert knowledge into land suitability analysis. In: *Value of LIS/GIS: Real and Perceived GIS Technology and its Relevance to Real World Problems Facing Your Organization. Vol. II.* Madison, University of Wisconsin Press, 24–37.

Williams, T.H., 1985, Implementing LESA on a geographic information system – a case study. *Photogrammetric Engineering and Remote Sensing*, **5**, 1923–1932.

Wolock, D.M. and Price C.V., 1994, Effects of digital elevation model map scale and data resolution on a topography-based watershed model. *Water Resources Research*, **30**, 3041–3052.

Wood, S.R. and Dent, F.J., 1983, *LECS: A land evaluation computer system.* AGOF/INS/78/006 Vols. 5 and 6. Bogor, Indonesia, Ministry of Agriculture, Government of Indonesia,.

Wright, L.E., Zitzmann, W., Young, K. and Googins, R., 1983, LESA-agricultural land evaluation and site assessment. *Journal of Soil and Water Conservation*, **38**, 82–86.

Zhang, W. and Montgomery, D.R., 1994, Digital elevation model grid size, landscape representation and hydrologic simulations. *Water Resources Research*, **30**, 1019–1028.

Environmental modelling: issues and discussion

A.K. Bregt, A.K. Skidmore and G. Nieuwenhuis

12.1 INTRODUCTION

In this final chapter, problems related to the successful use of GIS and RS in the domain of environmental modelling are discussed. The basis for this discussion was a joint problem analysis by the participants and lecturers of the ITC workshop on 'Environmental modelling using GIS and remote sensing' using a so-called meta-plan procedure (please see the Preface for names and affiliations of participants and lecturers). Before presenting the joint problem analysis, a generic framework for dealing with geo-related questions in environmental management was presented to the participants. The results from the joint problem analysis are discussed below.

12.2 GEO-INFORMATION RELATED QUESTIONS IN ENVIRONMENTAL MANAGEMENT

Environmental management, generally, requires three different kinds of information. Firstly, the status of particular environmental indicators or variables must be estimated. For instance, where are the most polluted sites? Where are the most and least productive soils? What is the land use in a region? Such 'what is where' questions are answered by individual surveys or observations; globally, the number of these surveys are enormous.

The second type of geo-information of interest to researchers and land managers is change over time. Typical questions may include: what is the change of the forested area in southeast Asia? How many hectares of agricultural land were abandoned in western Europe since 1990? What is the change in air pollution in the United States in the last 10 years? The 'what is changing where' questions are answered by multiple observations over a time period. Such monitoring may involve periodically recording certain parameters at selected spots, or, if one is interested in area changes, a full survey at regular time intervals may be necessary.

Thirdly, managers are interested in projections, both spatially and temporally; such projections may be characterized by 'what will be where' questions. For example: what will happen with the environmental pollution if we implement measure A? What will be the effect of urbanization on nature development in densely populated areas? Of the three types of questions, the 'what will be where' type is most difficult to answer. Besides an initial state of the land use (the 'what is where' question), an understanding of the process of change as well as the interaction between factors is required. This is extremely difficult, especially when biophysical and social-economical processes are considered together. In practice we see that it impossible to answer the question *'what will be where'* with great precision. Instead of a crisp prediction, we see the development of scenario's that represents possible lines of development.

In Table 12.1 the requirements in terms of data and processes of the three types of questions are indicated.

Table 12.1: Requirements for different questions.

Question	Data	Models	Techniques
What is where?	Spatial data	Model for spatial data	Standard DBMS and GIS
What is changing where?	Spatial data at time intervals	Model for space-time data	Standard DBMS and GIS
What will be where?	Spatial data at time intervals	Model for space-time data Model for projection	GIS, modelling environments and frameworks

Geo-spatial tools required for the analyses in Table 12.1 may be defined. For example, the *'what is where'* question requires a geographic information system (GIS) to store spatial data and make simple (Boolean) enquiries. The main constraints with this type of analysis lie in the domain of data access and data accuracy. For the *'what is changed where'* questions, multiple observations of land cover are required. Remote sensing techniques seem attractive for monitoring, but inaccurate classification, especially when compounded over time, may lead to unacceptably low accuracies. The analysis techniques available in GIS for analyzing spatial time series are also limited (Bregt and Bulens 1998). Methods need to be developed to survey, manage and analyze spatial data over time. Finally, for the *'what will be where'* questions, a combination of data, representing the initial status, and some rules or models describing the change of the environment over time, are needed. These rules range from relatively simple expert tables describing change in discrete intervals over time to complex dynamic simulation models describing change at continuous time intervals (see Chapter 2). It should be noted that the more complex dynamic simulation models are usually deductive, for example describing physical or chemical processes, such as the transport of water or the acidification of the soil (see Chapters 2 and 9). For exploring the *'what will be where'* question, data and transformation rules are combined in what is often called a spatial decision support system (SDSS) (see also Chapter 11). These systems generally consist of a standard GIS with some additional software components to facilitate decision support.

12.3 PROBLEMS RAISED BY THE PARTICIPANTS

At the end of the course, participants and lecturers discussed problems encountered in the practice of environmental management using GIS and remote sensing. This discussion was structured using a so-called meta-plan procedure. According to this procedure all the participants were asked to write, on separate pieces of paper, problems related to the topic of the course. Subsequently, all the problems raised were discussed and grouped into four main categories. In total 56 problems were mentioned. In Table 12.2 the main categories are presented.

Table 12.2: Main problems in the field of environmental modelling.

Main problem	Times mentioned	Percentage
Data	27	48
Modelling	13	23
GIS and RS technology	10	18
Expertise	6	11

Participants mentioned the issue of data for environmental modelling as being most problematic (see Chapters 3 and 4). The technology and the lack of expertise are considered to be minor problems, perhaps indicating that the technology and the development of appropriate skills are reaching a stage of maturity. Let us now have a closer look at the issues within the main problem categories.

12.3.1 Data problems

Within the data problem category the following issues were mentioned (note that the number of responses for each issue is in brackets):

- Lack of data and gaps in data coverage (spatial and temporal) (8)
- Accuracy and error propagation (7)
- Data costs (3)
- Data incompatibility (2)
- Data sharing
- Lack of common format of data
- Data collection
- Implementation of remote sensing observations in models.

Many practical problems are encountered with data in GIS modelling. Firstly, there is a lack of data, while data sets (spatial and temporal) frequently have gaps. If data are available, then accuracy may be insufficient to answer questions. Secondly, the cost of data is high reflecting the fact that field observations and mapping are time consuming and therefore expensive. Thirdly, there is hardly any programme focused on the systematic collection of spatial data to monitor the earth system; this problem is compounded by data from different sources often being incompatible (e.g. different scales, different classification systems being used in legends etc.), a problem discussed in Chapter 8. Remote sensing observations cannot be directly used in models. To convert radiation intensities into relevant parameters requires additional research. In general, data are scattered, incomplete, of variable quality and poorly documented.

12.3.2 Modelling problems

Within the modelling problem category the following issues were mentioned during the meta-plan procedure:

- Merging of data; scaling problems (5)
- Validation of models (2)
- Models: not good enough
- Incorporation of dynamic models in GIS
- Schematization of the complex real world
- Too much modelling and too little analytical work
- Developing models.

The participants and lecturers have clearly considered modelling in its widest sense; both the modelling of dynamic spatial processes as well as data modelling *per se*. The modelling and integration of data collected at different scales is considered to be a most important problem (see also Section 12.3.1). The validation of models is mentioned twice – there is clearly overlap with the problem of the quality of input data mentioned in the data problem section above. It was felt that the models are still not good enough, for example in the areas of dynamic modelling and dealing with the real world. During the course, and this book, it is shown that environmental models are useful for scenario studies on the state of the environment, and numerous models have been demonstrated (for example, see Chapter 2). The correct use of models requires, however, high quality models, in-depth knowledge of the model(s) being used, as well as systematically collected input data.

12.3.3 GIS and remote sensing technology problems

Within the technology problem domain the following issues were mentioned:

- Handling of remote sensing images from new sensors (e.g. hyperspectral, high spatial resolution, radar) (4)
- Remote sensing image interpretation (3)
- User-friendly GIS and remote sensing packages and specific software tools (2).

There are commercial systems available for GIS and remote sensing image processing as well as low budget systems with a remarkable functionality. Nevertheless, remote sensing image processing still requires specific software and hardware to be able to process data over large areas and to handle the new types of remote sensing data (see Chapter 3). The participants felt that remote sensing image interpretation is still a problem; this perhaps partly reflects the fact that the course focussed on GIS modelling, and not remote sensing *per se*. However, it is clear that the software for handling remote sensing images requires specific knowledge and is perceived as not being user-friendly.

12.3.4 Expertise problems

Within the expertise problem domain the following issues were mentioned:

- Integrated knowledge about GIS, RS and modelling
- GIS/RS; too far from user
- Cross-disciplinary work
- Practical know-how
- Lack of expertise
- Lack of trained persons.

Within the broad problem field of 'expertise', no single issue emerged. Topics mentioned may be broadly grouped as how to combine GIS and remote sensing technologies, enable inter-disciplinary work, and overcome a shortage of expertise.

12.4   PROPOSED SOLUTIONS FOR PROBLEMS BY PARTICIPANTS

The participants were then asked to indicate solutions for the problems mentioned in Section 12.3. The same procedure was followed: individual solutions written on paper were submitted to the session coordinators, and this was followed by a discussion among the participants and lecturers.

In total, 30 solutions were suggested. In Table 12.3, the groupings under the main problem categories are presented. A new category (General) was added for solutions that did not apparently fall under one of the main problem categories.

Table 12.3: Number of solutions for the main problem categories.

Main problem	Number of solutions	Percentage
Data	15	50
Modelling	5	17
GIS and RS technology	3	10
Expertise	3	10
General	4	13

Most of the solutions presented by the participants were given for the data problem (50 per cent). Let us now have a closer look at the suggested solutions.

12.4.1 Data solutions

The following solutions were suggested for the data problem (note that the number of responses for each issue is in brackets):

- Improved spatial and temporal resolution of RS (4)
- Standardization of spatial data (2)
- Lower the cost of data (2)
- Global climate database
- Global data accessibility network
- Development of data infrastructures and metadata
- Data warehouse
- More databases
- Common data format
- More ground observations.

The number of satellite systems is increasing, especially with commercial operators entering the market (see Chapter 3). The participants and lecturers expected high resolution imagery to be most useful for environmental modelling. Most of the other solutions suggest improved methods to store and distribute spatial data, for instance by using standardized databases that are easily accessible via the Internet. The development of spatial data infrastructures and easy accessible (global) spatial databases at low cost were considered to be the main solutions for the data problem.

12.4.2 Modelling solutions

The participants and lecturers suggested that the following ideas and developments would assist in solving some of the problems inherent in models:

- Better translation of models to applications
- More focussed models which deal with specific applications
- Development of new models
- Improved image processing models to insert remote sensing data into GIS
- Increased use of GIS in environmental and resource modelling and planning.

In summary, it was felt models that are better focused towards specific applications are required, both in the area of GIS as well as remote sensing.

12.4.3 GIS and RS technology solutions

It was noted above (Section 12.3.3) that technology was not considered a major constraint by the participants and lecturers. Indeed, the focus was on the integration of data from new remote sensing systems into GIS models. The following two points were made by the course participants

- More research on remote sensing data quality improvement (2)
- New sensors.

Solutions were expected from the use of new sensors and further research in the field of remote sensing data quality. Surprisingly, no solutions were mentioned in the field of faster hardware, though better software was implicit in the comments for improved modelling solutions. Apparently, the current GIS and remote sensing technology is fast and user-friendly enough to support users.

12.4.4 Expertise solutions

The following solutions were suggested to relieve the expertise problem:

- More GIS and remote sensing education
- Practical courses in GIS and RS
- Inter-disciplinary work.

A simple conclusion is that the main solution to the expertise problem is education, with an emphasis on a multi-disciplinary approach.

12.4.5 General solutions

We grouped the following solutions presented by the participants and lecturers under a 'general' category as no consistent theme was apparent:

- Increased public awareness
- Development of applications with users
- Better products for end users
- Mobile GIS/RS/GPS systems (hand-held).

Perhaps the general solutions can be best summarized as improved communication. The communication with end users and the general public was considered an important missing element in GIS modelling. The need to develop portable mobile systems may also be perceived as a solution to the communication problem.

12.5 REFLECTION

Perhaps the most interesting result is that the problems perceived with GIS modelling for environmental applications, that is data and models, has not changed in the last decade. At the First International Conference on Integrating GIS and Environmental Modelling held in Boulder Colorado, during September 1991, a similar list of problems emerged (Crane and Goodchild 1993). This summary highlighted systems issues (i.e. open architecture for model development, as well as systems benchmarking), data issues, GIS tools issues (i.e. tools fell short of what users require), and finally that a taxonomic catalogue of environmental models was required in order to better understand each model's value and its ability to interface with GIS (Crane and Goodchild 1993). In contrast, 15 years ago the main problem perceived in implementing GIS was hardware (Croswell and Clark 1988).

The lack of progress could be because our scientific knowledge of the Earth system accumulates at an agonizingly slow pace (Crane and Goodchild 1993), or perhaps it is because as human nature strives for knowledge, the perception is that progress is glacial. Certainly there is continuing progress in many areas of GIS applied to environmental modelling. We hope this book contributes in a small way to knowledge, and stimulates further research, demonstration and operation of GIS for environmental modelling.

12.6 REFERENCES

Bregt, A.K. and Bulens J.D., 1998, Integrating GIS and process models for land resource planning. In Heineke, H.J., Eckelmann, W., Thomasson, A.J., Jones, R.J.A., Montanarella, L.B., Buckley, B. (eds.). Land information systems: developments for planning the sustainable use of land resources. European Soil Bureau Research Report no. 4, 293-304.

Crane, M.P. and Goodchild,M.F., 1993, Epilog. In Goodchild, M.F., Parks, B.O., Steyart, L.T. (eds.). *Environmental modeling with GIS*. New York, Oxford University Press, 481-483.

Croswell, P.L. and Clark, S.R., 1988, Trends in automated mapping and geographic system hardware. *Photogrammetric Engineering and Remote Sensing*, **54**, 1571-1576.

Index